GENETIC APPROACHES TO
DEVELOPMENTAL NEUROBIOLOGY

GENETIC APPROACHES TO DEVELOPMENTAL NEUROBIOLOGY

Edited by
YASUZO TSUKADA

With 118 Figures and 38 Tables

UNIVERSITY OF TOKYO PRESS
SPRINGER-VERLAG BERLIN HEIDELBERG NEW YORK
1982

Dr. Yasuzo Tsukada
Professor, Department of Physiology,
School of Medicine, Keio University, Tokyo

Sole distribution rights outside Japan granted to
Springer-Verlag Berlin Heidelberg New York

ISBN 978-3-642-49325-6   ISBN 978-3-642-49323-2 (eBook)
DOI 10.1007/978-3-642-49323-2

Originally published by
UNIVERSITY OF TOKYO PRESS

2131/3140–543210

Contents

III. MAMMALIAN MUTANTS

Executive Committee

Foreword

The development of modern medicine has contributed to clarifying the etiology and treatment of various diseases as well as to improving public health and welfare, and basic biological research has discovered a great deal about the underlying mechanisms at work. However, there are still many diseases which remain mysterious and incurable, and biological processes which are still imperfectly understood. In order to promote research in medicine and biology, non-governmental funding, as well as government support, is very important. The Japan Medical Research Foundation has been in existence since October 1973 in order to meet such needs with aid from non-governmental financial sources. One of the Foundation's main activities is to hold annual symposia and seminars on various diseases and research problems.

To elucidate the mechanisms of the development of the nervous system in mammals is one of the most challenging problems for researchers in the field of developmental biology. Are there any formulae in the development of this complicated system? What factors are involved genetically and environmentally? Only a few questions have been answered so far.

However complicated the mammalian nervous system is, it develops and differentiates from one fertilized egg. Therefore, studies on the early development of mammals and the genetics of mammalian embryos will give us basic knowledge for further research. Experiments using mutant animals will also be necessary for such studies, combining genetics and development. At the symposium on genetic approaches to developmental neurobiology, researchers from different countries exchanged their knowledge and research results. I hope it will be a stimulation for further research on the development of the mammalian nervous system.

The Japan Medical Research Foundation is very pleased to sponsor this symposium, and I am hopeful that the publication of its proceedings will make its work available to a wide scientific readership.

June 10, 1982

Masayoshi Yamamoto
President
Japan Medical Research Foundation

ix

Preface

It is well known that the development of the nervous system is controlled by genetic and environmental factors, but showing the roles of these factors neurobiologically is still an urgent problem.

Thus a symposium concerned with the genetic control of the organogenesis and differentiation of the nervous system and the publication of its proceedings is an important occasion in the field of neurobiology.

Such an international symposium, on Genetic Approaches to Developmental Neurobiology, was held in Tokyo on May 11–12, 1981, under the auspices of the Japan Medical Research Foundation.

The symposium focused on mammals and consisted of three parts: Early Development of Mammals, Genetics of Mammalian Embryos, and Mammalian Mutants. The symposium was an instructive opportunity for all participants to discuss a new field of neurobiology. One of the most important findings presented showed that experiments using chimeric animals from neurological mutants are valuable in solving the pathogenesis of neurological diseases. The role of the gene itself on the development and differentiation on the nervous system, however, is still a future problem.

Both the symposium and the publication of the proceedings were made possible by the cooperation of the participants and the sponsoring organizations.

I wish to express my sincere appreciation to all members of the organizing and executive committees and the participants. I would also like to express my deep gratitude to the Japan Medical Research Foundation for promoting and sponsoring this symposium and publication.

June 17, 1982

Yasuzo Tsukada, M.D.
Chairman of the Organizing Committee

Opening Remarks

It is my great honor and pleasure to give the opening address to this symposium entitled Genetic Approaches to Developmental Neurobiology. On behalf of the Organizing Committee, I express our sincere gratitude to all of you for your participation, especially to those who have traveled from abroad so far to join us.

The symposium is supported by the Japan Medical Research Foundation, which was founded in 1973 for the purpose of promoting research in the medical field, particularly concerning intractable diseases—the so-called *nanbyo*.

During the past decade, research on neurobiology has greatly progressed. However, genetic approaches to this field are still immature. Therefore, the aim of this symposium is to understand the organogenesis of the nervous system, particularly the early development which is under genetic control. Several new techniques for this investigation have been developed, such as application of monoclonal antibody to the identification of specific macromolecules and making chimeric animals.

The symposium is constructed of two parts: one is the early development of mammalian embryos, and the other is the problems of chimeras and neurological mutants.

I believe that the new information and ideas derived from this symposium will contribute to our understanding of nervous function. Also, the fundamental knowledge of genetic control over the developing nervous system might lead to new gene therapy technology for treating hereditary neurological diseases in the future.

This symposium will cover only parts of the genetic approaches to neurobiology; it might be considered a trial run for the promotion of developmental neurobiology.

I would like to conclude my address by expressing our sincere appreciation to the Japan Medical Research Foundation for its generous sponsorship of the symposium, and to all of you participating in this meeting.

I hope that the meeting will be fruitful and stimulating for us all.

May 11, 1981

Yasuzo Tsukada, M.D.

I. EARLY DEVELOPMENT OF MAMMALS

Patterns of Cell Proliferation
in the Preimplantation Mammalian Embryo

Yoshihiro Kato, Kimie Yamazaki, Shiari Kimura,
Michiko Hayasaka, and Shunzo Kondo

INTRODUCTION

Cell proliferation plays a fundamental role during the preimplantation period of mammalian development.[1] In this study, several aspects of the proliferative process were examined in two species of rodents, in the mouse, and in the Chinese hamster.

The preimplantation development of the Chinese hamster was investigated with special reference to its cell proliferation patterns. Also, the developmental features of the mouse and the Chinese hamster were compared and analyzed with respect to cell proliferation. Analyses were made on the cell cycles at the first cleavage and during the blastocyst formation of a mouse embryo.

TIMING OF MAIN EVENTS DURING THE PREIMPLANTATION PERIOD OF THE CHINESE HAMSTER

Cell cultures from the Chinese hamster, *Cricetulus griseus*, have been used extensively for cell studies because of the unusual ease of identification of individual chromosomes in this species ($2n = 22$). There is, however, little precise information in the literature concerning the developmental processes of this potentially useful animal.[2] We have attempted to make a normal staging of the preimplantation development of the Chinese hamster with special attention given to cell proliferation.[3]

We have made microscopic observations on living specimens, as well as on the squashed and sectioned eggs and embryos obtained from the naturally mated females. The observations described here are based on approximately 1,500 eggs and embryos, collected at one- or two-hour inter-

Laboratory of Developmental Biology, Mitsubishi-Kasei Institute of Life Sciences, Minamiooya, Machida, Tokyo, Japan

3

vals from ovulaton to implantation. The breeding condition adopted in this study is listed in Table 1.

The preimplantation development of the Chinese hamster occurs within approximately 126 hours, during which time the following main events may be recognized in sequence: ovulation, fertilization, 1-cell (pronucleate), 2-cell, 4-cell, 8-cell, the appearance of the inside cell, 16-cell (early blastocysts), blastocyst formation, shedding of the zona pellucida, and implantalion.

The matured oocytes in the ovary reach metaphase of the second meiotic division just before ovulation and remain in this stage during ovulation. The ovulated eggs are surrounded by an egg membrane, the zona pellucida, and numerous cumulus cells. The diameter of the ovulated egg is approximately 95μm. Fertilization takes place in the upper end of the oviduct (ampulla). After the release of the second polar body, the penetrated sperm head transforms into the male pronucleus, which can be distinguished from the female pronucleus by its larger size and greater distance from the polar bodies.

Cleavage is total and nearly synchronous up to the 4th division. There is no clear-cut state of compaction at the 8-cell stage, and the first appearance of the inside cell (future inner cell mass, ICM) is noted at the 9-cell stage. The blastocoel formation takes place at approximately the 10-cell stage to lead embryos to the blastocyst stage. There is no stage of expansion, as is seen in the mouse blastocyst. After the shedding of the

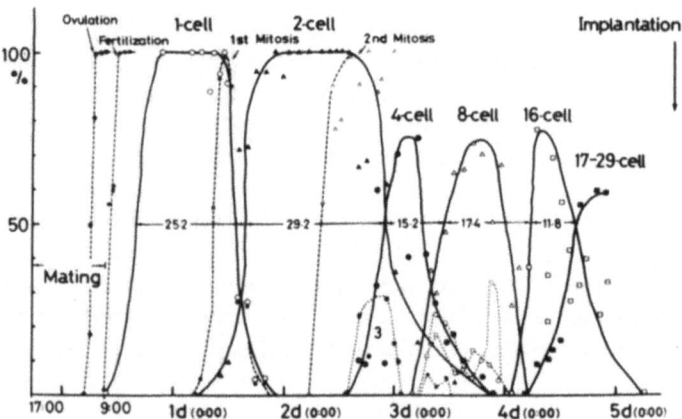

FIG. 1.   Percentages of the major events during the preimplantation stage of the Chinese hamster. Abscissa: time after mating.[3]

zona pellucida, the blastocysts attach themselves to the uterine wall to initiate an early phase of implantation. The total cell number of the blastocyst at this stage is approximately 45.

By expressing the frequencies of the appearance of given phases (% basis), along the axis of developmental time elapsed after ovulation, the timing and sequence of the main events leading to implantation can be determined (Fig. 1). A high degree of synchrony during cleavage is notable. The approximate timing of the main events under the present breeding conditions is shown in Table 1.

TABLE 1. Timing of the Events leading to Implantation[3]
(Chinese hamster)

Stage	Day*
Mating starts	−1day, 5p.m.
Ovulation	0day, 6:15a.m.−7:20a.m.
Mating ends	0day, 9a.m.
Fertilization	0day, 7:00a.m.−1p.m.
One-cell (Pronucleate + Mitosis)	0day, 1p.m.−1day, 2p.m.
2-cell	1day, 2p.m.−2day, 7p.m.
4-cell	2day, 7p.m.−3day, 10a.m.
8-cell	3day, 10a.m.−4day, 4a.m.
16-cell (Early blastocyst)	4day, 4a.m.−4p.m.
Blastocyst	4day, 4p.m.
Shedding of zone pellucida starts	4day, 10p.m.
Onset of implantation	5day, 12a.m.

Light control: 6a.m.−10p.m. (light)/10p.m.−6a.m. (dark)
* 0 day: The day when mating ended.

COMPARISON OF PREIMPLANTATION DEVELOPMENT IN THE MOUSE AND THE CHINESE HAMSTER

The most characteristic feature of the Chinese hamster during its preimplantation period is the low cell number necessary for an embryo to reach the stage of implantation. Approximately 45 cells are necessary for a blastocyst to attach itself to the uterine epithelium. In contrast, approximately 100–200 cells are present in the mouse blastocyst at the time of implantation.[4] Figure 2 shows a comparison of the increase in cell number during the preimplantation period in the mouse (ICR) and the Chinese hamster.

The preimplantation periods in the mouse and Chinese hamster last about four days and four and a half days, respectively. Seven successive

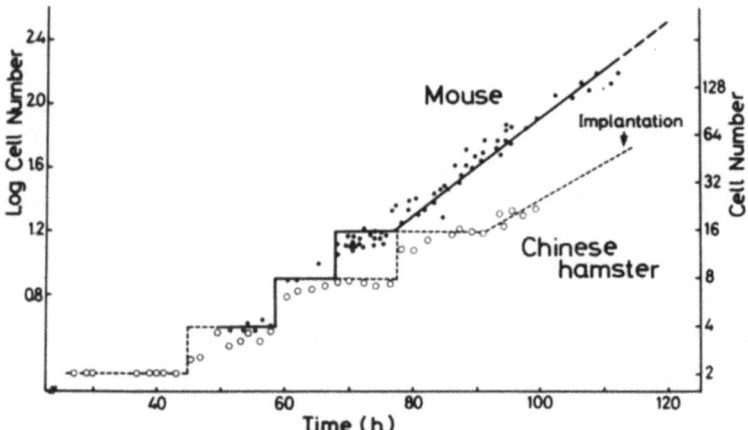

FIG. 2. Comparison of the increase in cell number between the Chinese hamster and the mouse during the preimplantation period.

cell divisions occur in the mouse, whereas only five successive divisions occur in the Chinese hamster during preimplantation development. A variety of differentiative and morphogenetic events have been reported to be dependent upon cell divisions and cell cycles. Comparison of several preimplantation events in the mouse and the Chinese hamster was made to determine if the difference in cell proliferation patterns and rates might be correlated with other developmental events.[5]

Differentiation of nucleolus

Nucleolar structure offers a good measure for the degree of differentiation of early mammalian eggs. As development progress in the mouse, the nucleolar structure at an ultrastructural level develops from a dense, fibrilar matrix to a granular element.[6-8] Figure 3 compares the nucleolar structures of the blastomeres of the Chinese hamster with those of the mouse, both at the 8-cell stage. In the Chinese hamster, which reaches the stage of implantation with fewer divisions (five divisions for the Chinese hamster and seven for the mouse), the differentiation of the nucleolus is more advanced and therefore is granular in structure.

The first appearance of the inside cell

A geometric packing of blastomeres of equal size which would allow one to be entirely surrounded by others requires 13 blastomeres.[9] The

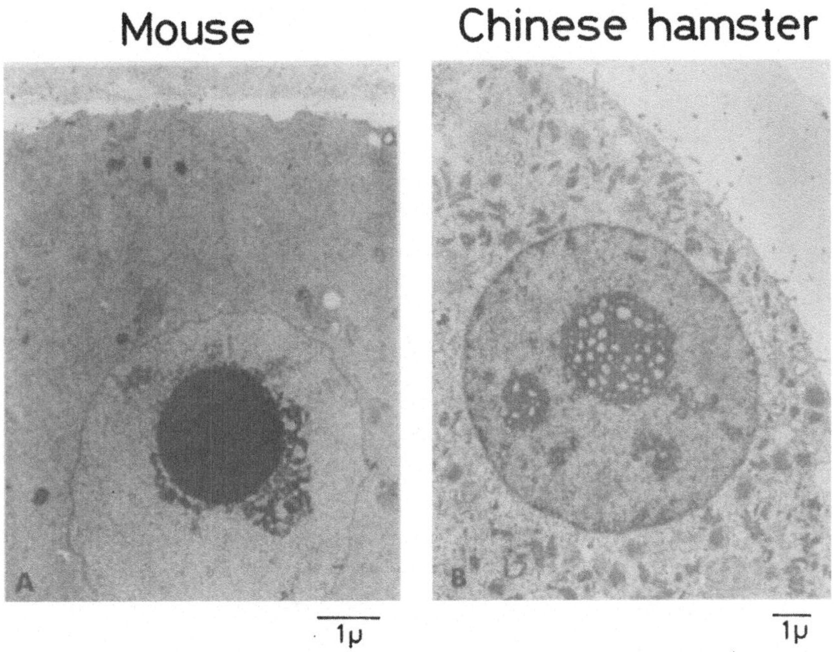

Mouse Chinese hamster

FIG. 3. Comparison of the nucleolar structures of the mouse and the Chinese hamster at the 8-cell stage.[5]

process of cleavage should lead an egg to arrange at least one blastomere to be positioned inside. By reconstructing the serial sections of eggs at different developmental stages, the total cell number at the time of the first appearance of the inside cell can be determined. Figure 4 shows drawings from the eggs of the Chinese hamster and the mouse. The minimum number of blastomeres needed to have at least one cell inside is 13 for the mouse and 9 for the Chinese hamster. Such a considerable difference in total cell number at the time of appearance of the first inside cell is compatible with the different degrees of nucleolar differentiation found in the two species.

Blastocoel formation

McLaren[4] points out that six visible events take place during the preimplantation period of a mouse in the following sequence: 1. cleavage; 2. compaction; 3. junction formation; 4. vesicle release; 5. appearance of blastocoel; and 6. ICM and trophectoderm differentiation. In the mouse,

Chinese hamster Mouse

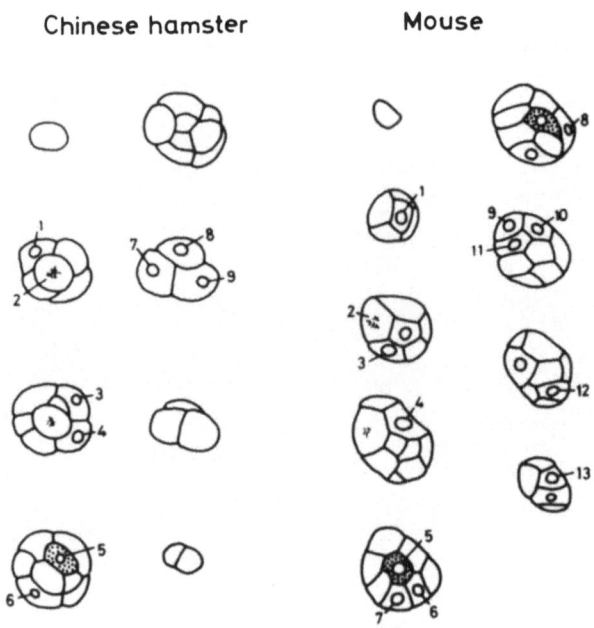

FIG. 4. The first appearance of the inside cell in the eggs of the Chinese hamster and the mouse. Individual blastomeres are numbered. Dotted blastomeres are the first inside cells to appear.

it has been reported that there is a sequential relationship among compaction, junction formation, vesicle release, and the formation of the blastocoel.[10]

Compaction at the 8- to 16-cell stage in the mouse results from the changes in cell shape and the peripheral junction formation.[11] Blastocoel formation seems to be dependent both on compaction and on the release of fluid from the cytoplasmic vesicles into the intercellular spaces.[4]

The four events listed above as prerequisites for the formation of the blastocoel in the mouse were examined with both light and electron microscopy in the Chinese hamster at the 8- and 10-cell stage. As mentioned earlier, no apparent phase of compaction was noted. The appearance of the blastocoel was not accompanied by drastic changes in cell shape. Blastomeres retained nearly the original spherical shape (Fig. 5). The number of cytoplasmic vesicles was much lower in the Chinese hamster than in the mouse blastomeres (8-cell stage), and these vesicles were randomly

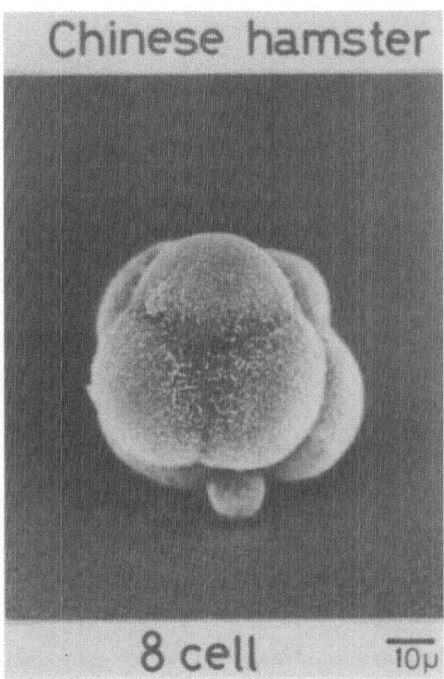

FIG. 5. Scanning electronmicrograph of an egg of the Chinese hamster at the 8-cell stage. No apparent compaction is seen ($\times$ 670).

distributed without showing any regional localization within the blastomeres (Fig. 6). Approximately the same number of tight junctions were found at the 8-cell stage in the mouse and in the Chinese hamster (Fig. 7).

These observations indicate that, in the eggs of the Chinese hamster, the cytoplasmic vesicle release, changes of cell shape and compaction do not seem to be correlated with the formation of the blastocoel. Apparently, the eggs and embryos of the Chinese hamster go through the preimplantation events successfully with a very small number of blastomeres in a somewhat different way than those of the mouse.

THE CELL CYCLE ANALYSES IN THE PREIMPLANTATION DEVELOPMENT OF THE MOUSE

The cell cycles during embryonic development in a variety of species

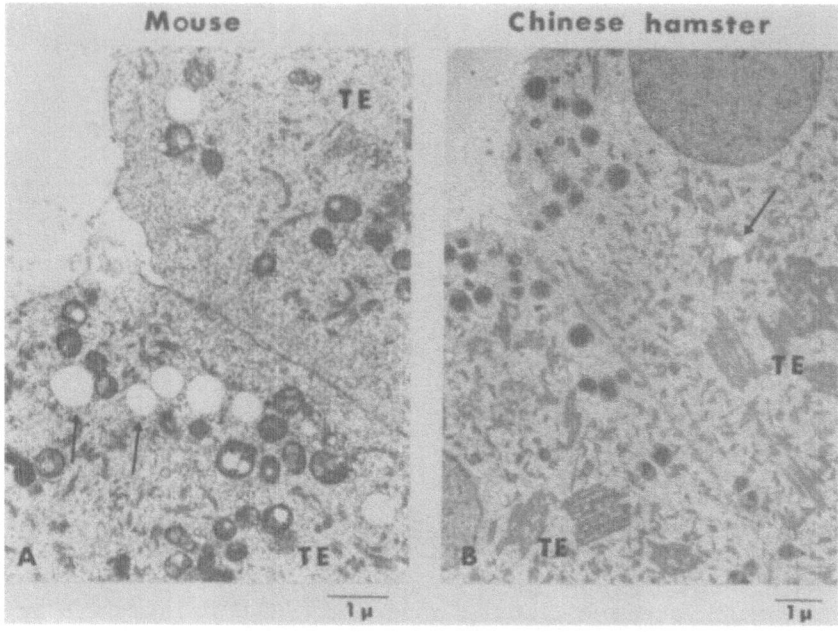

FIG. 6. Comparison of the cytoplasmic vesicles between the blastomeres of the mouse and the Chinese hamster at the 8-cell stage. In the mouse, abundant vesicles (indicated by arrows) are located at the interface of the blastomeres ($\times$ 8,000). In the Chinese hamster, vesicles (indicated by an arrow) are few and are distributed randomly ($\times$ 6,150). TE: trophectoderm.[5]

are quite different from those of adults. In the early development of the frog[12] and the sea urchin,[13,14] for instance, total generation time (T_C) is usually very short, the G_1 and G_2 phases are lacking, and the duration of the DNA synthetic phase (T_S) is also short. As development proceeds, the G_1 and G_2 phases appear and T_S and T_C lengthen.[1] A statistical analysis has shown that a definitive correlation does exist between the developmental changes of a given cell cycle parameter and the total generation time.[15]

The cell cycles in the preimplantation stages of mammals seem to be exceptional, since the G_1 and G_2 phases are present from the very beginning of development.[16,17] Shortening, rather than lengthening, of T_C and T_S takes place after implantation.[18] Since obtaining precise infor-

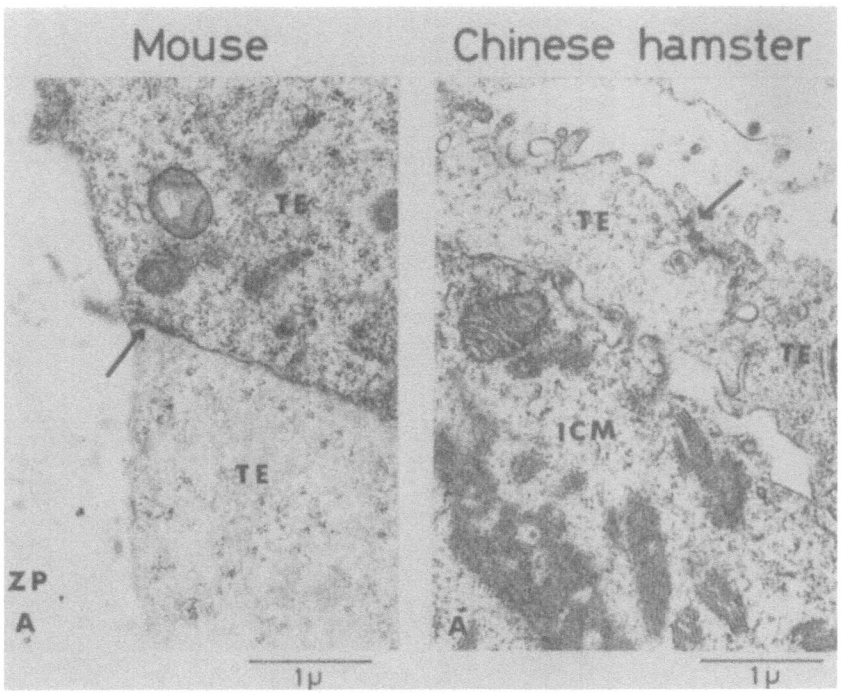

FIG. 7.   Electronmicrograph of the tight junction in the blastomeres of the mouse and the Chinese hamster, at the 8-cell
stage ($\times$ 16,000). TE: trophectoderm, ICM: inner cell mass.[5]

mation about the cell cycle is essential for an exact understanding of cell
proliferation, we have performed several lines of studies on the cell cycle
during the preimplantation period of the mouse embryo (ICR/SLC).[19]

*Re-evaluation of the cell cycle determination at the first cleavage in the mouse
embryo*

Table 2 summarizes the results obtained by other investigators on the
cell cycle at the first cleavage of a mouse embryo.[16,17,20−22] In all of these
studies, actual determinations of the mitotic phase are lacking. A direct
determination of the duration of mitosis (T_M) is essential, if one wants to
obtain accurate values for the cell cycle phases. We have re-examined the
cell cycle at the first cleavage in the ICR/SLC mouse. Figure 8 shows our
criteria for determining mitotic phases.

The first cleavage was found to be highly synchronous (Fig. 9). Eggs

TABLE 2. Cell Cycle Time in Hours (1st Cleavage) (Mouse)

Mouse	T_C	T_{G1}	T_S	T_{G2}	T_M	Method	Investigators
HS	Not measured	Absent	Not measured	12*		³H-TdR pulse, *in vitro*	1970. Gamow *et al.*
CFl	Not measured	≒1	6	12	Not measured	³H-TdR pulse, *in vitro*	1975. Luthardt *et al.*
ICR/Ha	12	Not measurable	7	4	1**	³H-TdR pulse, *in vitro*	1976. Mukherjee
Swiss mice	24.1	1.3	6.1	15.4	1.3**	Densitometry, *in vivo*	1978. Sawicki *et al.*
Inbred	19	0.5	7	11.5*		Cytofluorometry, *in vivo*	1980. Streffer *et al.*

*: $T_{G2} + T_M$, **: $T_C - (T_{G1} + T_S + T_{G2})$

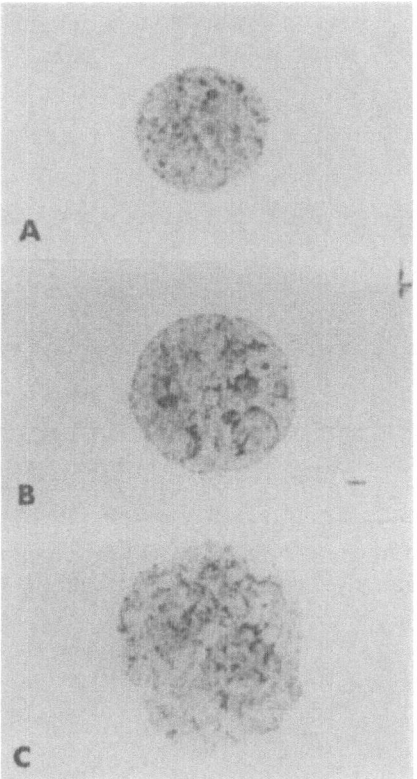

FIG. 8. Nuclear phases of the mouse blastomeres. Fixed
with Carnoy's solution and stained with Giemsa. A: telophase
at the 1st cleavage. No nucleolei are seen. B: interphase at
the 2-cell stage. Distinct nucleolei. C: prophase at the 2nd
cleavage. Dispersion of nucleolei and condensation of chromatin ($\times$ 640).[19]

were collected at one-hour intervals from 28 hours after HCG treatment
until the second cleavage began. The collected eggs were pulse-labeled with
^{3}H-TdR (10 μCi/ml; specific activity: 25Ci/m mol) for 15 minutes at 37° C
under 5% CO_2 in air. Autoradiographs were prepared and the frequency of
the appearance of a given cycle phase was followed along the time axis of
development (Fig. 10). Table 3 shows the durations of individual cycle
phases and T_C thus obtained. No G_1 phase was noted and a relatively
long T_M was found.

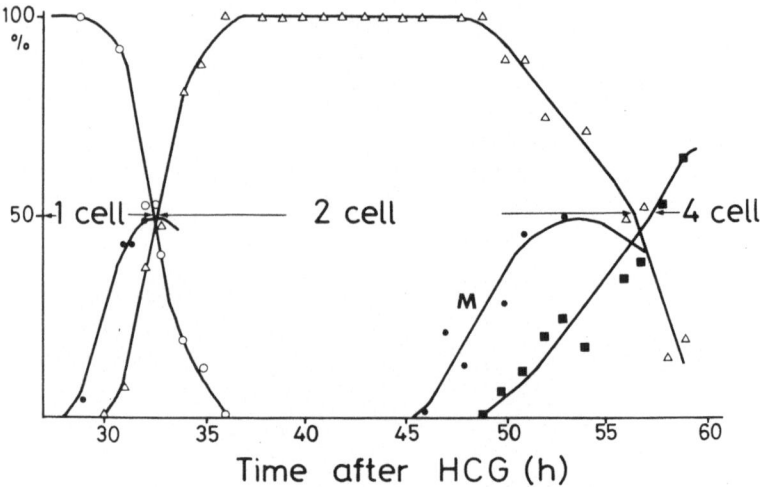

FIG. 9.   The degree of synchrony in the mouse eggs at the 1st cleavage. Abscissa: time after HCG treatment (h). Ordinate: % cells at a given cleavage stage. M: mitosis.[19]

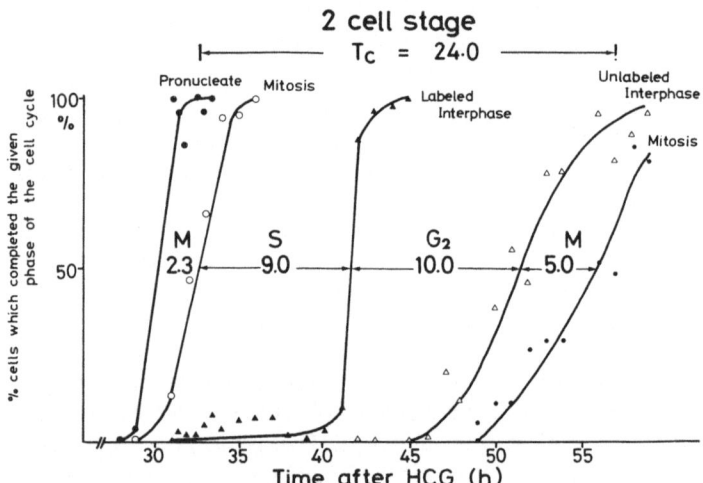

FIG. 10.   Transit of cells throughout the pronucleate to the 2nd mitoses (left to right). Abscissa: time after HCG treatment (h). Ordinate: % cells which have completed a given phase of the cell cycle.[19]

TABLE 3. Cell Cycle Time in Hours (1st Cleavage)[19]
(Mouse)

T_C	T_{G1}	T_S	T_{G2}	T_M
24	$\doteq 0$	9.0	10.0	5.0

An analysis of the cell cycle in mouse blastocysts

In the blastocyst of a mammalian embryo, the first cellular differentiation takes place. The blastocyst consists of two cell populations: the inner cell mass (ICM) that forms the embryo proper and the outer trophectoderm cells that subsequently differentiate into extraembryonic derivatives.

Barlow et al.[23] concluded that, in mouse blastocysts, the ICM cells were dividing faster than the trophectoderm because the former had a higher labeling index. Their conclusion was based on two assumptions: (1) the proportion of the cell cycle occupied by the S phase remains the same in two regions, and (2) the S phase occupies the same position relative to the two cell cycles. Consequently, the labeling index will not directly reflect the rate of cell proliferation.

We have studied cell proliferation and the cell cycle in the ICM and trophectoderm of mouse (ICR strain) blastocysts by means of cell counting on whole mounts and by microdensitometry for nuclear DNA determinations.[24] To distinguish the trophectoderm from the ICM in whole mount preparations, an immunosurgical technique was used. When blastocysts were treated first with anti-mouse antisera and then with a Guinea-pig complement, the trophectoderm, which occupies the outside position, was selectively killed.[25]

We have found that, during the process of immunosurgery, only the nuclei of trophectoderm cells showed drastic nuclear condensation, the ICM cells, in contrast, retained their normal nuclear size (Fig. 11). By determining the number of nuclei, the DNA content of individual nuclei, and the mitotic indices, it was possible to calculate the doubling time of cells belonging to ICM, trophectoderm, or the blastocyst as a whole (equivalent to the length of the cell generation cycle, T_C). Using this information, the cell cycle phases of the ICM and the trophectoderm were also determined.

A comparison of the cell division rate of the two cell populations shows that the trophectoderm cells grow more rapidly than the ICM cells (Fig. 12), contrary to previous reports. The doubling times of the trophectoderm and ICM cells at the 32–64-cell stage (early blastocyst) were 8.5 h and 11h,

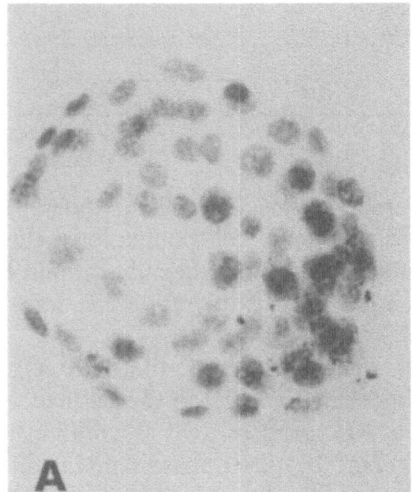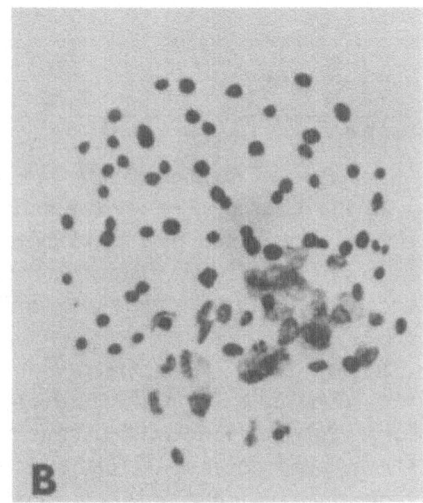

FIG. 11.   Condensation of nuclei of the trophectoderm cells of a mouse blastocyst by immunosurgical treatment. A: before treatment. B: after treatment. Only the nuclei of the trophectoderm cells are condensed. Feulgen staining ($\times$ 320).[24]

respectively. At the 64–128-cell stage (late blastocyst), the doubling times were 6.5 h for the trophectoderm and 13 h for the ICM. The durations of the cell cycle phases in the trophectoderm and ICM at the 32–64-cell stage are shown in Table 4. For T_{G1}, T_{G2}, and T_M, the durations of individual phases were almost identical in the trophectoderm and the ICM. On the other hand, there was a considerable difference in the length of T_S between the two cell populations. Based on these observations, we conclude that the rapid cell growth in the trophectoderm is the result of its shorter T_S as compared with that of the ICM.

When an embryo enters the stage of blastocyst formation, a number of differences appear between ICM and trophectoderm. Such diversifications have been observed in labeling indices, radiation sensitivity, enzyme activities (e.g. alkaline phosphatase), and the electrophoretic patterns of tissue-specific polypeptides.[26] The present study has added yet another such difference, namely, the asynchrony of the cell cycle phases in the two cell populations of a mouse blastocyst.

A number of investigations on early mouse embryos have revealed that cell death takes place during preimplantation development.[27,28] The effect of cell death on the determinations of cell proliferation does not seem to

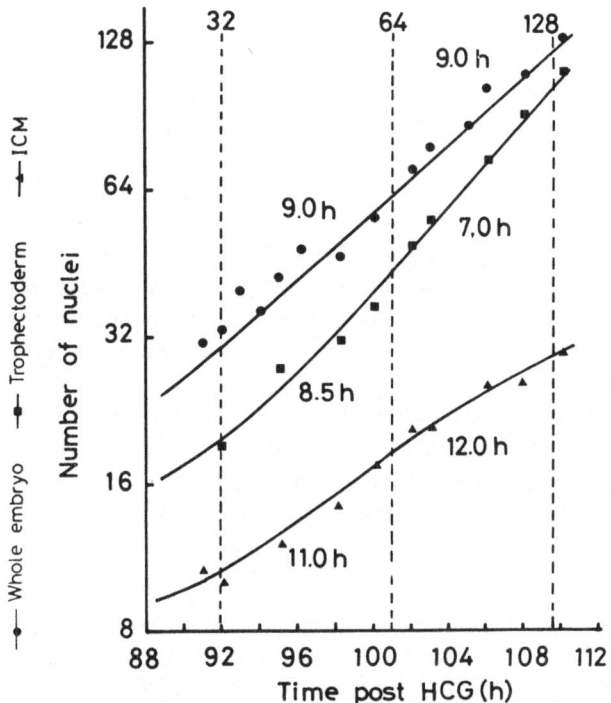

FIG. 12. The relation of cell number of the development of mouse blastocysts. Abscissa: time after HCG treatment. Ordinate: number of nuclei in the ICM (▲), trophectoderm (■) and a whole embryo (●).[24]

TABLE 4. Length of the Cell Cycle and of the G_1, S, G_2 and M Phases (h)[24] (Mouse)

Cell stages	Phases	Whole embryo	ICM	Trophectoderm
32–64	Tc	9.0	11.0	8.5
	G_1	2.0	2.0	2.3
	S	4.6	6.7	3.7
	G_2	1.8	1.7	1.8
	M	0.6	0.6	0.7

be considerable, at least at the blastocyst stage where the cell cycle determination was made in this study. Approximately 1.24% of the cells are reported to die at the 32–64-cell stage.[28] We omitted the apparently dead cells at the time of cell counting and cell cycle determinations.

CONCLUSIONS

The cell cycle and cell proliferative patterns during the preimplantation development of mammals are unusual and complicated in many ways. In contrast to other developmental systems, the cell cycle time is relatively long and a long duration of the G_2 phase is also noted during early cleavage. The shortening of the cell cycle is reported to take place relatively late, only after implantation.[18]

The fact that the preimplantation embryos develop in the maternal reproductive tracts presents various methodological difficulties when cell cycle determination is attempted. Differences due to the species and strain are also frequently found.[1] The occurrence of cell death further complicates the study.[27,28]

Despite such technical difficulties, increasing interest in early mammalian development and technical improvements in cell cycle studies have been noticeable and invite further investigations, since cell proliferation and differentiative events are mutually interrelated.[29]

REFERENCES

1. Graham, C. F.: The cell cycle during mammalian development. In: The Cell Cycle in Development and Differentiation (ed. M. Balls and F. S. Billett), p. 293. Cambridge University Press, Cambridge, 1973.
2. Pickworth, S., Yerganian, G. and Chang, M. C.: Fertilization and early development in the Chinese hamster, *Cricetulus griseus. Anat. Rec.*, **162**: 197–208, 1969.
3. Kato, Y. and Yamazaki, K.: In preparation.
4. McLaren, A.: Early events in mammalian embryogenesis. In: Mechanisms of Cell Change (ed. J. D. Ebert and T. S. Okada), p. 49. John Wiley & Sons, Inc., New York, 1977.
5. Kato Y., Yamazaki, K. and Kondo, S.: In preparation.
6. Calarco, P. G. and Brown, E. A.: An ultrastructural and cytological study of the preimplantation development in the mouse. *J. Exp. Zool.*, **162**: 253–284, 1969.
7. Hillman, N. and Tasca, R. J.: Ultrastructural and autoradiographic studies of mouse cleavage stage. *Am. J. Anat.*, **126**: 151–174, 1969.
8. Van Blerkom, J. and Runner, M. N.: The fine structural development of preimplantation mouse parthenotes. *J. Exp. Zool.*, **196**: 113–123, 1976.
9. Izquierdo, L.: Cleavage and differentiation. In: Development in Mammals (ed. M. H. Johnson), Vol. 2, p. 99. North-Holland, Amsterdam, 1977.
10. Smith, R. and McLaren, A.: Factors affecting the time of formation of the mouse blastocoele. *J. Embryol. exp. Morph.*, **48**: 37–51, 1978.
11. Ducibella, T.: Surface changes of the developing trophoblast cell. In: De-

velopment in Mammals (ed. M. H. Johnson), Vol. 1, p. 5. North-Holland, Amsterdam, 1977.

12. Graham, C. F. and Morgan, R. M.: Changes in the cell cycle during the early embryonic development of *Xenopus laevis*. *Devel. Biol.*, **14**: 439–460, 1966.

13. Hinegardner, R. T., Rao, B. and Feldman, D. E.: The DNA synthetic period during early development of the sea urchin egg. *Exp. Cell Res.*, **36**: 53–61, 1964.

14. Dan, K., Tanaka, S., Yamazaki, K. and Kato, Y.: Cell cycle study up to the time of hatching in the embryos of the sea urchin, *Hemicentrotus pulcherrimus*. *Develop., Growth & Differ.*, **22(3)**: 589–598, 1980.

15. Kato, Y.: Profile analysis of the cell-cycle parameters in scale development. *Develop., Growth & Differ.*, **17(3)**: 293–294, 1975.

16. Luthardt, F. W. and Donahue, R. P.: DNA synthesis in developing two-cell mouse embryos. *Devel. Biol.*, **44**: 210–216, 1975.

17. Sawicki, W., Abramczuk, J. and Blaton, O.: DNA synthesis in the second and third cell cycle of mouse preimplantation development. A cytophotometric study. *Exp. Cell Res.*, **112**: 199–205, 1978.

18. Snow, M. H. L.: Embryo growth during the immediate postimplantation period. In: Embryogenesis in Mammals. Ciba Foundation Symposium 40, p. 53. Elsevier/North-Holland, Amsterdam, 1976.

19. Kato, Y. and Yamazaki, K.: In preparation.

20. Gamow, E. I. and Prescott, D. M.: The cell life cycle during early embryogenesis of the mouse. *Exp. Cell Res.*, **59**: 117–123, 1970.

21. Mukherjee, A. B.: Cell cycle analysis and X-chromosome inactivation in the developing mouse. *Proc. Nat. Acad. Sci. U.S.A.*, **73(5)**: 1608–1611, 1976.

22. Streffer, C., Van Beuningen, D., Molls, M., Zamboglou, N. and Schulz, S.: Kinetics of cell proliferation in the preimplanted mouse embryo *in vivo* and *in vitro*. *Cell Tissue Kinet.*, **13**: 135–143, 1980.

23. Barlow, P., Owen, A. J. and Graham, C.: DNA synthesis in the preimplantation mouse embryo. *J. Embryol. exp. Morph.*, **27**: 431–445, 1972.

24. Kimura, S. and Kato, Y.: In preparation.

25. Solter, D. and Knowles, B.: Immunosurgery of mouse blastocyst. *Proc. Nat. Acad. Sci. U.S.A.*, **72**: 5099–5102, 1975.

26. Johnson, M. H.: Intrinsic and extrinsic factors in preimplantation development. *J. Reprod. Fert.*, **55**: 255–265, 1979.

27. Handyside, A. H.: Time of commitment of inside cells isolated from preimplantation mouse embryo. *J. Embryol. exp. Morph.*, **45**: 37–53, 1978.

28. Copp, J.: Interaction between inner cell mass and trophectoderm of the mouse blastocyst. I. A study of cellular proliferation. *J. Embryol. exp. Morph.*, **48**: 109–125, 1978.

29. Gardner, R. L.: The relationship between cell lineage and differentiation in the early mouse embryo. In: Genetic Mosaics and Cell Differentiation (ed. W. J. Gehring), p. 205. Springer-Verlag, Berlin, 1978.

The Analysis of Cell Lineages in the Postimplantation Mammalian Embryo

Rosa S. P. Beddington

INTRODUCTION

Cell lineages have been studied extensively in the preimplantation mouse embryo.[1-5] These studies have done much to illuminate the stages at which the fate of cells becomes restricted during the earliest steps of differentiation. For example, no invariant lineage can be traced through cleavage which might explain the initial differentiation of trophectoderm and inner cell mass (ICM) cells, the two primary tissues of the blastocyst.[4] Instead this choice is thought to be influenced by environmental or positional cues as opposed to ancestral ones.[6,7] However, once the trophectoderm and ICM have differentiated to form a well expanded blastocyst, they appear to constitute two separate lineages, cells from one tissue being unable to give rise to the other either in the embryo[8-10] or *in vitro*.[11,12] Although there is some controversy regarding the precise timing of this divergence[13-18] the weight of the evidence suggests that it is complete by the mid-blastocyst stage. A similar binary division occurs shortly afterwards with the emergence of the embryonic ectoderm and primitive endoderm, both of which are derived from the ICM.[19,20] Once again, these two populations behave as discrete lineages, apparently committed to mutually exclusive developmental pathways.[2,5,12,21] These experiments emphasize two things. Firstly, they stress the binary nature of cell commitment during preimplantation differentiation: cells can form either trophectoderm or ICM; either embryonic ectoderm or primitive endoderm. Secondly, it appears that once commitment has occurred, this is a heritable property, stably maintained within a particular lineage. This is pleasantly clear-cut. One can *describe* preimplantation development in terms of relatively simple rules. However, this should not be confused with an understanding of the *mechanisms* specifying particular differentiation steps. For example, although there is a strong correlation between certain developmental pathways and traceable lineages there is no evidence that ancestry *per se* is instrumental in

Sir William Dunn School of Pathology, South Parks Road, Oxford, England.

specifying a particular developmental pathway. The relative importance of cellular environment and cell lineage has yet to be resolved for any differentiation event in the mouse. Nonetheless, the current knowledge of cell lineages and cell fate in the preimplantation mouse embryo provides the essential framework on which to base a sensible molecular dissection of early developmental events in the mammal.

The differentiation occurring in preimplantation embryogenesis appears to be preoccupied with the necessary adaptations for viviparity, the segregation of prospective extraembryonic tissues and membranes.[22] These events shed no light on the origins and organization of the fetal primordia. The emergence of the mammalian fetus occurs after implantation and, therefore, poses severe problems for a direct analysis of cell lineages and cell fate. For this reason, relatively little is known about the early differentiation of the fetal tissues. This is obviously a serious deficiency, not only in terms of understanding normal embryogenesis but also with regard to recognizing factors which may contribute to abnormal development. Therefore, the object of this chapter is to review some of the work which has attempted to elucidate cell lineage relationships relevant to the initial organization of the fetus in rodent embryos.

THE GROSS MORPHOLOGY OF EARLY FETAL ORGANIZATION

The emergence and initial organization of the fetus is intimately associated with the process of gastrulation, an event which does not begin until almost a third of the way through gestation in the mouse.[23] There is strong evidence (see below) that all the fetal primordia are descended from a single population of cells, the embryonic ectoderm, and that gastrulation is the process whereby this epithelial sheet of cells is converted into a highly complex structure composed of a variety of tissue types and embodying the basic design of the fetus. Gastrulation, therefore, represents the key to fetal organization.

The process of gastrulation begins on the 7th day of gestation, in the mouse, with the appearance of the primitive streak.[23,24] Before this, the embryonic region of the egg cylinder consists only of an inner layer of embryonic ectoderm and an outer, investing layer of primitive endoderm (Fig. 1). The primitive streak arises in the posterior part of the embryonic ectoderm and defines unequivocally the antero-posterior axis of the embryo. Embryonic ectoderm cells invaginate through the streak and differentiate into mesoderm. Subsequently, the mesoderm spreads as two lateral sheets between the ectoderm and endoderm to form a third inter-

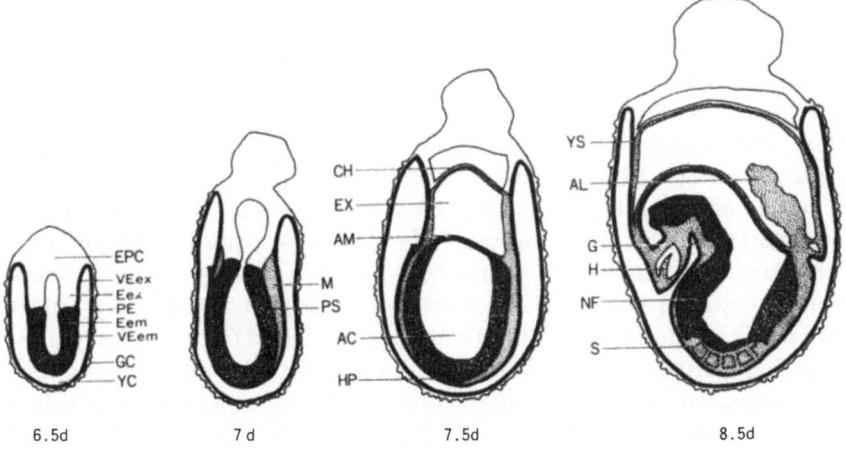

FIG. 1. A diagrammatic illustration of the morphological changes occurring during gastrulation and the first stages of organogenesis. EPC, ectoplacental cone; VE_{ex}, extraembryonic visceral endoderm; E_{ex}, extraembryonic ectoderm; PE, parietal endoderm; E_{em}, embryonic ectoderm; VE_{em}, embryonic visceral endoderm; GC, giant cells; YC, yolk cavity; M, mesoderm; PS, primitive streak; CH, chorion; EX, exocoelom; AM, amnion; AC, amniotic cavity; HP, head process; VYS, visceral yolk sac; AL, allantois; G, gut; H, heart; NF, neural fold; S, somite.

mediate layer in the embryonic region (Fig. 1). Some mesoderm also moves into the extraembryonic region where it makes a major contribution to the fetal membranes: the chorion, the amnion, and the visceral yolk sac. It also gives rise to the allantois (Fig. 1). At the anterior end of the primitive streak a group of morphologically distinct mesoderm cells are delaminated. These constitute the head process which consists of tightly packed cells extending anteriorly along the midline, apparently displacing the outer layer of primitive endoderm.[25,26]

Organogenesis commences before gastrulation is complete. A comprehensive account of organogenesis is not appropriate here and a more detailed description can be found elsewhere.[23,24] However, some of the salient features of the earliest stages of organ formation will be outlined in order to illustrate both the complexity and the rapidity of the changes that take place (Fig. 1). Late on the 8th day of gestation the headfold appears at the anterior end of the embryo and the ectoderm lying along the antero-posterior axis thickens to form the neural plate. Subsequently, this rolls up into the neural tube, the direct precursor of the brain and spinal cord. Beneath the headfold the heart primordium emerges and by the end of the 9th day this may already show signs of rhythmical contraction. The mesoderm is subdivided into the notochord (a narrow tube-like structure) running beneath the neurectoderm, rows of somites (segmented blocks of mesoderm) on each side of it and, more peripherally, lateral plate mesoderm. Blood vessels appear within the lateral plate mesoderm and a circulatory system is established between the embryo and the visceral yolk sac by the 10th day. The visceral yolk sac is probably the major nutritive organ of the conceptus at this stage. It is derived from the extraembryonic mesoderm together with an investing layer of visceral extraembryonic endoderm, and surrounds the amnion, being separated from it by a greatly enlarged exocoelom (Fig. 1). The gut forms beneath the notochord, appearing as a closed tube at the anterior and posterior ends of the embryo whilst, initially, the midgut region remains open to the yolk cavity. The mesoderm at the posterior end of the embryo proliferates to form the allantois which grows in a mesometrial direction through the exocoelom to fuse with the chorion and later contribute to the formation of the placenta. Therefore, within approximately 2.5 days a single epithelial layer has been transformed from a simple cylindrical sheet into a complex, highly organized entity showing most of the features, albeit rudimentary, characteristic of the mouse fetus.

LINEAGE STUDIES ON THE POSTIMPLANTATION RODENT EMBRYO

At the moment neither direct observation nor precise manipulation of the early postimplantation rodent embryo are possible *in utero*. Therefore, the lineage relationships between tissues in the egg cylinder and the various organs of the later conceptus have been inferred mostly from indirect analyses. Three particular problems will be examined in relation to early fetal organization:

    i) Which of the early embryonic tissues contribute to the fetus itself?

    ii) When do particular fetal primordia segregate?

    iii) Do prospective fetal primordia show strict topographical relationships to one another during gastrulation?

These questions will be considered in the context of the various *experimental* strategies which have been adopted to study cell lineages in the postimplantation embryo. Descriptive studies on gastrulation and the initial organization of the fetus, of which there are several,[25,26] will not be discussed separately but only mentioned where they provide corroboration of experimental results.

I. Manipulation of the preimplantation rodent embryo

The analysis of cell fate and cell potency in the mouse blastocyst has clearly delineated the early progenitors of the various tissue layers in the postimplantation egg cylinder. If ICM cells from the expanded blastocyst are injected into genetically distinct blastocysts[28] or aggregated with morulae[9,10] their progeny can be detected only in the fetus and fetal membranes and not in the ectoplacental cone or giant cells. Similarly, if blastocysts are reconstituted from genetically distinct ICMs and trophectoderm vesicles only the fetus and fetal membranes, with the exception of the chorion, are derived entirely from the ICM.[19,20] Furthermore, in the few cases where rat ICMs from separate egg cylinders within mouse trophectoderm analysis of the postimplantation embryo reveals that the ectoplacental cone, extraembryonic ectoderm and trophoblast giant cells are always of mouse origin.[19] Extraembryonic ectoderm will participate in subsequent development after being injected into blastocysts and it has been shown to colonize only the ectoplacental cone, the chorion and giant cells.[21] These results provide powerful evidence for the trophectodermal origin of the extraembryonic ectoderm, ectoplacental cone and trophoblast giant cells and show that the embryonic region of the egg cylinder is derived from the ICM.

The production of preimplantation chimeras has also indicated that only one tissue, the embryonic ectoderm, is responsible for generating all the tissues of the fetus. The injection of single 5th day embryonic ectoderm cells into the blastocyst results in chimerism throughout the fetus including the gut and its derivatives. In addition, the amnion and the mesodermal component of the visceral yolk sac are colonized.[5] Furthermore, some of these chimeras showed colonization of the germ line as well as somatic tissues.[29] In contrast to these results, single primitive endoderm cells from the same stage, distinguished by the rough appearance of their surface, contributed progeny only to the endoderm layer of the visceral yolk sac[5] and to the parietal endoderm.[30] Similarly, primitive endoderm from the egg cylinder on the 6th day of gestation, when returned to the blastocyst, colonizes only the visceral yolk sac endoderm[20] and the parietal endoderm.[30] This indicates that the primitive endoderm, formed on the 5th day, is not the ancestor of definitive gut endoderm, the latter arising some time after the 6th day of gestation from the embryonic ectoderm. The normal fate of the visceral endoderm in the embryonic region of the egg cylinder is unknown. It is not clear whether these cells die *in situ,* to make way for the definitive gut endoderm, or whether they migrate to the extraembryonic region and contribute to the formation of the visceral yolk sac. Unfortunately, embryonic ectoderm from the postimplantation embryo fails to colonize the blastocyst[31] although it has been reported that unspecified cells from the 8th day embryo will, at a very low frequency, continue to proliferate and form fetal chimeras following injection into the blastocyst.[32,33] Therefore, the evidence from chimeras points to embryonic ectoderm being the sole founder tissue of the fetus. However, these studies have not shed any light on the manner in which the embryonic ectoderm cells of the *postimplantation* embryo become allocated to particular developmental pathways, associated with the formation of the different fetal primordia.

II. Ectopic transfer of postimplantation tissues

An alternative method for examining the developmental potential of postimplantation tissues is to assess their histogenic potential in ectopic sites.[34] There is, of course, no reason to suppose that the differentiation of embryonic cells in such a site should coincide with their normal fate, a criticism which may also be applied to asynchronous or heterotopic injections into the mouse blastocyst. However, the analysis of ectopic grafts is well suited to detect restrictions in histogenic potential which presumably reflect similar restrictions occurring in the embryo. Apart from very rare, if not heroic exceptions,[35] this sort of *in vivo* assay of histogenesis has been confined to the study of tissues as opposed to single cells.

The embryonic ectoderm from pre- and early primitive streak stages in the rat will continue to develop beneath the kidney capsule and give rise to teratomas, benign tumors containing a chaotic array of differentiated tissues representative of all three germ layers of the fetus.[36] Similar results are obtained from mouse embryonic ectoderm isolated at an equivalent stage, although in some cases these grafts retain an additional population of undifferentiated, proliferating cells,[37] so-called embryonal carcinoma cells (EC cells), which are classified as malignant stem cells.[38] From the diverse array of mature tissues seen in ectopic grafts one may deduce that, as a tissue, embryonic ectoderm remains pluripotent at least until the later stages of gastrulation. Furthermore, in the rat, no difference was found between teratomas derived from the proximal as opposed to the distal half of the embryonic ectoderm,[39] although a profound difference between these two types of graft might not be anticipated since both will contain fractions of the anterior embryonic ectoderm and the primitive streak. So far no detailed study has been undertaken to look for regionalization in the embryonic ectoderm with respect to its histogenic potential.

In both the rat and the mouse no mature tissues are observed in grafts of the visceral embryonic endoderm isolated from the gastrulating embryo. In rats these grafts are resorbed[36] whereas in the mouse they are either resorbed or give rise to parietal endoderm-like cells.[37] These results are entirely consistent with those from chimera experiments and reinforce the argument that it is the embryonic ectoderm of the early gastrula which serves as the precursor of all adult tissues.

It seems likely that definitive endoderm does not appear in the embryo until the headfold stage. It is at this stage that grafts of embryonic ectoderm, while still capable of forming mesodermal tissues, fail to generate gut derivatives in ectopic sites.[40] Endoderm from this stage does not develop in isolation, and grafts of mesoderm alone produce growths composed almost entirely of brown adipose tissue. However, if endoderm and mesoderm are transferred together then teratomas containing mesodermal and gut derivatives, but no definitive ectoderm tissues, are produced.[40] Furthermore, regional differences with respect to the type of gut derivatives formed are immediately evident. Grafts consisting of all three germ layers from the anterior part of the rat embryo (the neural plate region) give rise to foregut derivatives. The posterior part (the primitive streak region) differentiates into derivatives characteristic of the midgut and hindgut. This regionalization is retained even if the ectoderm is removed before grafting, indicating that it is a property of either the mesoderm or the endoderm or, perhaps, dependent on an interaction between the two.[41] Such overt regionalization is less apparent in the mesodermal and ectodermal derivatives formed in the teratomas. However, the middle part of the embryo (Hensen's node region) does show a reduced tendency to generate well differentiated skin and the anterior part of the embryo produces teratomas deficient in brown adipose tissue.[41] It is possible that brown adipose tissue represents a true mesodermal product whereas the "mesodermal" derivatives found in anterior grafts may be formed predominantly from the neural crest.

Hence the analysis of experimental teratomas indicates that at the onset of gastrulation embryonic ectoderm has the ability to provide all the fetal precursors. However, towards the end of gastrulation, and certainly by the headfold stage, the potential of embryonic ectoderm has been restricted so that it forms only definitive ectoderm and mesoderm. It seems likely, therefore, that gut endoderm, originating from the embryonic ectoderm, is laid down before the headfold stage. It is interesting that, in the mouse, the ability of embryonic ectoderm to generate EC cells also appears to be lost by the headfold stage, again indicating a restriction in potential.[42]

The ability of embryonic ectoderm to generate EC cells may have relevance to the origin of the germ line in the mouse. It has been recognized for some time that EC cells and germ cells share many distinctive characteristics,[43-45] perhaps the most important of which is their apparent totipotency.[46,47] Furthermore, it has been shown that both spontaneous and experimental teratocarcinomas (those teratomas containing EC cells) can originate from the germ cells in mature and immature gonads.[48-50] Therefore, the origin of EC cells from the 7th day mouse embryo[36] could be due to the presence of a separate population of cells, the primordial germ cells, within the embryonic ectoderm. Certainly, histochemical staining has revealed a group of alkaline phosphatase positive cells, presumed to be the primordial germ cells, at the posterior aspect of the primitive streak in 8th day mouse embryos.[51] However, it has been claimed that the ectopic transfer of primitive streak stage mutant embryos (Sl^J/Sl^J and W/W), supposedly deficient in germ cells, results in the formation of teratocarcinomas.[51] Although the status of the germ cells in these mutant embryos during gastrulation has not been clearly defined[53] it has been argued that these transfer experiments indicate that EC cells can have a somatic origin.[52] The sex and strain specificities observed in germ cell derived teratocarcinomas, which are not applicable to those derived from the early embryo, also suggest that EC cells may have a dual origin.[54] If this is true and embryonic ectoderm can act as the immediate precursor of EC cells then this provides an example of a tissue which shows a restriction in developmental potential, in terms of its inability to generate either trophoblast or primitive endoderm, which can still give rise to totipotent cells. This would set a precedent for the embryonic ectoderm also being able to give rise to the germ cells. Thus, the characteristic totipotency of germ cells would have to be considered as a differentiated trait, acquired some time after the 8th day of gestation.[55] Such a postulated change in potency within the germ line is not all that unreasonable considering that female germ cells are thought to undergo reversible X-inactivation during their development.[56]

III. Isolation experiments in vitro

Complementary studies on the histogenic capacity of isolated postimplantation tissues have been undertaken *in vitro*[57-59] but, in general, the results of these experiments are less reproducible than those obtained from ectopic grafts. In culture the three-dimensional organization of the explants tends to deteriorate and perhaps as a result of this the range of differentiated tissues generated is not as great as that seen *in vivo*. If tissues which have been allowed to differentiate in culture are transferred to an ectopic site *in vivo* further differentiation may result. This indicates that the repertoire of tissues formed in culture is not a true reflection of developmental potential.[59] On the other hand, certain sorts of differentiation, such as the formation of cardiac muscle, occur more readily in culture than *in vivo*.[60] This illustrates one of the problems associated with histogenic analyses. The results are always subject to the conditions under which histogenesis is assessed and, therefore, not always readily applicable to the embryo. Indeed, it has been argued that an embryonic environment, as opposed to an ectopic or *in vitro* conditions, should encourage the greatest expression of potency.[61]

IV. Manipulation of postimplantation embryos in vitro

The development of reliable culture systems for postimplantation mouse embryos has lagged far behind the establishment of those for supporting preimplantation development.[62,63] Those systems concentrating on the culture of mouse blastocysts over the implantation period have attempted to rectify this deficit, but the postimplantation development of the embryos tends to be unreliable with a high proportion of *in vitro* egg cylinders failing to complete gastrulation and effect normal organogenesis.[64-67] The work of New and his colleagues has greatly advanced culture techniques for the development and maintenance of postimplantation rat embryos.[68] For example, rat embryos can develop successfully *in vitro* from the pre-primitive streak stage until the 30–40 somite stage, during which time gastrulation is completed and organ differentiation well advanced.[69] Using a similar roller culture technique to that described by New[70] apparently normal development has been obtained in mouse embryos explanted at the late primitive streak stage (8th day) and grown in culture for 36 hours until the 8–12 somite stage[71] and in embryos explanted at the early somite stage and maintained *in vitro* for the following 48 hours of organogenesis.[72] It has also been shown that both early and late primitive streak stage mouse embryos can develop in static cultures, on a par with embryos maintained *in vivo*, for at least 24 hours.[73]

These recent advances in postimplantation mammalian embryo culture provide a means for studying cell lineages in the embryo itself. For example, using a similar strategy to that employed to study cell potency in preimplantation development, marked or labeled cells may be injected into primitive streak stage[71] or early somite stage (A. J. Copp and R. S. P. Beddington, unpublished observations) embryos which are subsequently grown *in vitro*. The distribution of donor cells in these "*in vitro* chimeras" can then be analyzed according to the marker or label used. Although mouse embryos grow normally in culture for only a limited period (24–48 hours) and, therefore, any study of cell fate or cell potency will be necessarily short term, the changes occurring during and immediately after gastrulation are sufficiently rapid[23,24] for this approach to provide valuable information. In some respects the acute nature of these experiments is an advantage since tritiated thymidine ([³H]-thymidine) can be used as a label for donor cells and, hence, chimerism analyzed by autoradiography in histological sections.

Cell fate has been analyzed, following the orthotopic injection of labeled clumps of embryonic ectoderm cells, in the late primitive streak stage mouse embryo.[71] Two important conclusions can be drawn from this study. Firstly, it confirms that embryonic ectoderm can give rise to gut endoderm in the embryo. Orthotopic injection of embryonic ectoderm from the distal tip of the egg cylinder (Hensen's node region) generates gut chimeras (Table 1). In two of these chimeras the labeled cells in the gut were closely associated with labeled cells in the notochord (Figs. 2A and 2B), which supports a previous claim, based on a careful histological investigation of gastrulation, that the notochord and gut originate from a common precursor, the head process.[25] This fits nicely since the head process emerges from the most anterior aspect of the primitive streak, Hensen's node.[25]

The second conclusion which can be drawn from these results is that there is regionalization in the embryonic ectoderm with respect to devel-

TABLE 1.   The Distribution of Donor Cells in "*in vitro* Chimeras" following Orthotopic Injections

Injection site	Anterior										Distal											Posterior														
Chimaera number	1	2	3	4	5	6	7	8	9	10	1	2	3	4	5	6	7	8	9	10	11	1	2	3	4	5	6	7	8	9	10	11	12	13	14	15
Surface ectoderm	▨	▨																																		
Neurectoderm				▨	▨													▨																		
Loose mesoderm												▨	▨		▨	▨	▨					▨	▨	▨	▨									▨		
Blood vessels													▨									▨	▨											▨		
Somite												▨			▨	▨						▨														
Gut													▨		▨																					
Notochord												▨																								
Primitive streak															▨	▨																				
Allantois																													▨	▨	▨					▨

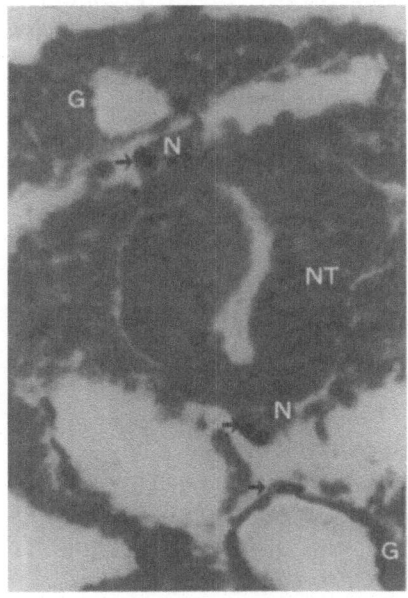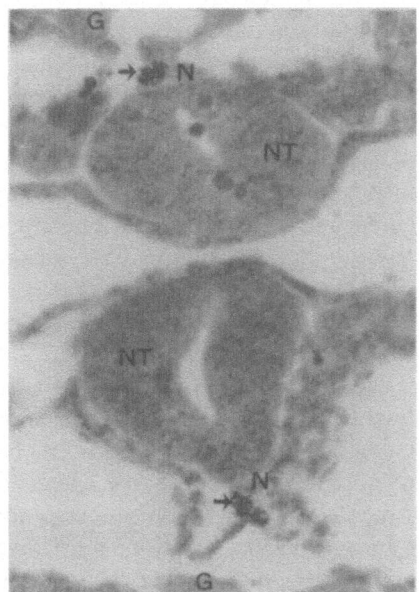

FIG. 2. A. Transverse section through chimera No. 3 (distal orthotopic injections). Donor cells are present in the notochord and gut.

B. Transverse section through the same chimera, at a different level, also showing the presence of donor cells in the notochord and gut.

NT, neural tube; N, notochord; G, gut. Donor tissue is marked by arrows.

opmental fate. Embryonic ectoderm injected orthotopically into the anterior region of the egg cylinder gives rise predominantly to derivatives of definitive ectoderm: neurectoderm and surface ectoderm (Table 1). In contrast, embryonic ectoderm from the distal tip of the egg cylinder very rarely contributes to definitive ectoderm tissues. Instead, it colonizes lateral plate mesoderm, somites, gut and notochord (Table 1). Finally, orthotopic injections into the posterior end of the primitive streak[74] produce chimeras which have been colonized exclusively in mesodermal tissues, both embryonic and extraembryonic (Table 1). It is, of course, not surprising that presumptive fetal primordia should show a predictable topographical arrangement. Fate maps have been described for many classes of embryo at the time of gastrulation.[75,76] However, it is reassuring that the somewhat crude topography described here for the presumptive tissues of the mouse is largely consistent with those fate maps, constructed at an equivalent stage of development, in avian embryos.[77,78]

It should be possible to determine, by heterotopic grafting, whether or not the predictable fate of embryonic ectoderm, in different regions of the late primitive streak stage egg cylinder, is due to the prior commitment of cells to a particular developmental pathway. In other words, if embryonic ectoderm from one site is placed in a different region of the embryo does it differentiate according to its site of origin, implying previous commitment, or will it conform to the developmental fate of its new location? Ideally, such an analysis should be clonal, because otherwise it is impossible to eliminate the influences of microenvironment in the tissue graft or the possibility of specific cell selection contributing to the final pattern of chimerism. This is one of the major limitations of *in vitro* chimeras. The short duration of the experiments precludes any sensible clonal analysis since clonal expansion will be minimal. Such an analysis, which is essential for an understanding of the segregation of cell lineages during fetal organization, must await a means of either prolonging the culture period or of producing chimeras *in utero*.

V. Teratology

A review of the various techniques which have been used to investigate cell lineage during gastrulation would be incomplete without a mention of teratological studies. These have illustrated the remarkable regulative capacity of the embryo before the onset of organogenesis. However, because the precise nature of the teratogenic insult is often unknown, and usually involves non-specific and random cell killing, these studies have not contributed information relevant to *specific* cell lineages.

If rat embryos are exposed *in utero* to either heat shock[79] or to X-irradiation[80] before the onset of mesoderm formation then any embryo which survives appears to have regulated completely and formed a normal fetus. However, if treatment is applied during gastrulation or the early stages of organogenesis then specific malformations ensue. The same pattern of susceptibility is seen in the mouse where screening for abnormalities in the skeletal system has revealed that irradiation up to the late primitive streak stage results either in death or complete regulation whereas, thereafter, specific vertebral abnormalities are induced.[81] A recent study, using Mitomycin C, which also causes extensive cell death, produces a similar effect when administered before and during gastrulation. Despite a loss of up to 85% of their cells those embryos which survive achieve almost complete regulation, although, for some obscure reason, post-natal mortality is extremely high.[82] Such efficient embryonic regulation suggests that when the drug has completed its cell killing pluripotent cells capable of replacing all the missing parts remain within the embryo. It is, of course, conceivable that individual cells are not pluripotent but that a sufficient number and variety of cells with complementary but restricted potentials persist to ensure reconstitution of the embryo, each part being replaced by the remnants of a specific precursor population of cells. This seems an unlikely explanation considering the vast reduction in cell number. One might expect whole populations of cells, responsible for the formation of a particular tissue, to be destroyed and hence particular deformities to arise. Apart from occasional micropthalmia and a general deficiency of germ cells such effects were not observed.[82]

Therefore, the application of teratogens has produced consistent results in the early postimplantation embryo. It is during or towards the end of gastrulation that malformations may be induced for the first time. It seems reasonable to conclude that up to this stage the embryonic ectoderm, the known precursor of the fetus, remains developmentally labile and can effect complete regulation. Thereafter, the organization of the prospective fetal tissues becomes more rigid and damage to any one part is more likely to have a pronounced and enduring effect. However, those defects occurring in any particular structure need not necessarily stem from direct damage to its primordia but may arise indirectly, perhaps due to deranged tissue interactions.

VI. The comparison of normal and mutant embryos

In some species mutant embryos have been used to clarify the normal processes of embryogenesis. For example, the study of the mutant *grand-childless* in *Drosophila* has helped to elucidate when the germ line originates in this insect.[83] Attempts have been made to use mutants in the mouse as a tool for dissecting developmental processes. The use of sterile mutants to investigate the origin of EC cells[52] has been mentioned earlier. The *T*-locus mutants in the mouse have also received much attention and it has been claimed that this locus is "a major region controlling the first determinative steps in the embryo."[84] This hypothesis has been further extended to explain the lethality of *t*-haplotypes on the basis of aberrant tissue interactions.[85] These claims have yet to be substantiated but, nonetheless, one of the mutations at the *T*-locus, t^{w18}, has been used to investigate embryonic development at the time of gastrulation.

Like other *t*-haplotypes, t^{w18} is a recessive embryonic lethal. It affects development at the primitive streak stage, homozygous embryos showing an abnormal overgrowth of the primitive streak and a marked deficiency in normal mesoderm.[85,86] When 9th day t^{w18}/t^{w18} embryos were transferred to the testis they produced teratomas, although at a very low frequency. These teratomas were unusual in that they contained only occasional traces of mesodermal derivatives and no tissues, such as skin or gut, whose formation depends upon mesodermal interactions.[87,88] Instead, the tumors consisted largely of rapidly proliferating neurectoderm and a few other definitive ectoderm derivatives such as squamous epithelial pearls. It has been proposed that this restricted expression of potency is due to the inability of mutant embryos to make the transition from embryonic ectoderm to mesoderm, while the ability to make definitive ectoderm is unaffected.

Subsequently, a comparison was made between the mitotic index in the primitive streak stage embryonic ectoderm of normal and t^{w18}/t^{w18} embryos.[89] The mutant embryos showed a disproportionately high mitotic index throughout the ectoderm but on close inspection it was found that the mitotic figures displayed an abnormal plane of cleavage. The only region where the mitotic index and the orientation appeared normal coincided with the location of the proliferative zone, a site identified in normal embryos at the anterior aspect of the primitive streak with an unusually high mitotic index.[90] A correlation has been drawn between the apparent normality of this zone in t^{w18}/t^{w18} embryos and their ability to undergo normal definitive ectoderm differentiation, the inference being that the proliferative zone in normal embryos is responsible for generating the definitive ectoderm of the fetus.[89,91] However, without further substantiation, such as testing the histogenic capacity of different regions of the embryonic ectoderm, and in particular that of the proliferative zone, this would seem rather an ambitious claim, not least because the characterization of the defect in t^{w18}/t^{w18} embryos cannot be considered as complete. Furthermore, the notion that the proliferative zone generates definitive ectoderm contradicts the results from *in vitro* chimeras, where embryos receiving orthotopic injections of embryonic ectoderm from the anterior aspect of the primitive streak (the distal tip of the egg cylinder) showed a distinct lack of definitive ectoderm colonization.[71] Therefore, so far, the use of mutant embryos as a tool for analyzing the onset of fetal organization has not proved particularly fruitful.

CONCLUSIONS

Six independent methods for investigating the earliest stages of fetal differentiation have been discussed. Although it is all too clear that very little is known about either cell lineages or tissue specification during gastrulation and organogenesis there are certain important consistencies which emerge from the results of these different studies.

i) *Which of the early embryonic tissues contribute to the fetus itself?*

All the available evidence points to the embryonic ectoderm being the sole founder tissue of the fetus. No other type of cell has been seen to colonize the fetus following injection into the blastocyst[22] and only embryonic ectoderm displays the requisite histogenic potential in ectopic sites.[36,37] Furthermore, postimplantation *in vitro* chimeras have shown that embryonic ectoderm can give rise to a wide variety of fetal tissues, including derivatives of all three germ layers, *in situ* in the embryo.[71] Therefore, it appears that the mouse embryonic ectoderm is, after all, analogous to the epiblast of the chick[78] and that until, and probably during the early part of gastrulation, this tissue has the ability to generate all the constituents of the fetus. In addition, circumstantial evidence suggests that the mouse embryonic ectoderm may also give rise to the germ cells and that there is no early segregation of the germ line in the mouse embryo.

ii) *When do particular fetal primordia segregate?*

As gastrulation progresses and organogenesis commences the increasing organization of the embryo appears to occur at the expense of developmental lability. This is demonstrated by the induction of specific defects in response to teratogens applied at the end of gastrulation, whereas earlier in development the embryo has the capacity for complete regulation.[79-82]

A restriction in potency can also be demonstrated in experimental teratomas. Embryonic ectoderm from the headfold stage has lost the ability to generate gut derivatives.[40] It is tempting to assume that once a founder population of definitive endoderm cells has been delaminated from the embryonic ectoderm that the gut endoderm then segregates as a self-contained lineage, in a manner similar to that seen in primitive endoderm formation.[5] However, such an assumption is obviously premature since there is no direct evidence that the cells of the definitive endoderm are confined to an exclusive developmental pathway. In fact, histological studies[25] and the distribution of labeled cells in *in vitro* chimeras[71] suggests that whichever cells give rise to the gut may also contribute to all or part of the notochord. However, like mesoderm, the differentiation of definitive endoderm involves a restriction in potential since neither of these tissues isolated from the fetus itself has been observed to form definitive ectoderm derivatives in ectopic sites.[39,40]

The production of mesoderm from embryonic ectoderm appears to be a more prolonged process, headfold stage ectoderm still showing the capacity to generate mesodermal derivatives.[40] However, it is not known at present whether mesoderm is a truly homogeneous tissue, all mesoderm, regardless of the stage or location of its formation, arising with equivalent developmental potential. The difficulty in distinguishing possible neural crest derivatives from those of "true" mesoderm in teratomas further confounds the issue. Therefore, it is impossible to say anything about the divergence and segregation of the various mesodermal derivatives in the embryo although *in vitro* chimeras could be used to investigate changes in potency occurring in this tissue. For example, one could test whether extraembryonic mesoderm and fetal mesoderm are immediately committed to mutually exclusive pathways or whether they remain interchangeable during the early stages of organogenesis.

iii) *Do prospective fetal primordia show strict topographical relationships to one another during gastrulation?*

Regional differences in developmental fate can be detected in the embryonic ectoderm at the late primitive streak stage.[71] It has been mentioned already that the topography of presumptive tissues in the mouse is similar to that described in chick embryos at an equivalent stage.[77,78] A more comprehensive comparison between the fate map in the chick and that charted for the gastrula of the urodele embryo has revealed considerable homology between the two.[75,78] As the organization of presumptive tissues in the mouse appears to conform to this basic pattern it seems reasonable to speculate that these three classes of vertebrate embryo share common principles of tissue allocation during gastrulation, despite the gross differences in their overall morphology. Furthermore, there is some evidence that this common pattern of cell fate may extend throughout the vertebrate phylum.[92]

It can be seen that recent work has clarified some aspects of mammalian gastrulation and early organogenesis but, compared with the detailed description of cell lineages in the preimplantation embryo, our understanding of tissue segregation during these later, crucial stages of development can only be considered as rudimentary. The problem is largely a technical one, which has confined analyses to that of tissues rather than single cells. The assessment of developmental potency in the blastocyst has illustrated the importance of clonal analysis in studying the relationship between cell lineage and cell fate. It seems clear that the basic rules for establishing fetal organization will become apparent only when a similar clonal approach can be applied to the postimplantation mouse embryo.

Acknowledgements

The author wishes to thank Professor R. L. Gardner and Dr. J. D. West for valuable discussion.

REFERENCES

1. Wilson, I. B., Bolton, E. and Cuttler, R. H.: Preimplantation differentiation in the mouse egg as revealed by microinjection of vital markers. *J. Embryol. exp. Morph.*, **27**: 467–479, 1972.
2. Gardner, R. L. and Papaioannou, V. E.: Differentiation in the trophectoderm and inner cell mass. In: The Early Development of Mammals (2nd Symposium of the British Society for Developmental Biology) (ed. M. Balls and A. E. Wild), pp. 107–132. Cambridge University Press, London, 1975.
3. Kelly, S. J., Mulnard, J. G. and Graham, C. F.: Cell division and cell allocation in early mouse development. *J. Embryol. exp. Morph.*, **48**: 37–51, 1978.
4. Graham, C. F. and Deussen, Z. A.: Features of cell lineage in preimplantation mouse development. *J. Embryol. exp. Morph.*, **48**: 53–72, 1978.
5. Gardner, R. L. and Rossant, J.: Investigation of the fate of 4.5 day *post coitum* mouse inner cell mass cells by blastocyst injection. *J. Embryol. exp. Morph.*, **52**: 141–152, 1979.
6. Tarkowski, A. K. and Wroblewska, J.: Development of blastomeres of mouse eggs isolated at the 4- and 8-cell stage. *J. Embryol. exp. Morph.*, **18**: 155–180, 1967.
7. Hillman, N., Sherman, M. I. and Graham, C. F.: The effect of spatial arrangement on cell determination during mouse development. *J. Embryol. exp. Morph.*, **28**: 263–278, 1972.
8. Gardner, R. L.: Origins and properties of trophoblast. In: Immunobiology of Trophoblast (ed. R. G. Edwards, C. W. S. Howe and M. H. Johnson), pp. 43–65. Cambridge University Press, London, 1975.
9. Rossant, J.: Investigation of the determinative state of the mouse inner cell mass, I: Aggregation of isolated inner cell masses with morulae. *J. Embryol. exp. Morph.*, **33**: 979–990, 1975.
10. Rossant, J.: Investigation of the determinative state of the mouse inner cell mass, II: The fate of isolated inner cell masses transferred to the oviduct. *J. Embryol. exp. Morph.*, **33**: 991–1001, 1975.
11. Ansell, J. D. and Snow, M. H. L.: The development of trophoblast *in vitro* from blastocysts containing varying amounts of inner cell mass. *J. Embryol. exp. Morph.*, **33**: 177–185, 1975.

12. Hogan, B. and Tilly, R: *In vitro* culture and differentiation of normal mouse blastocysts. *Nature* (Lond.), **265**: 626–629, 1977.

13. Handyside, A. H.: Time of commitment of inside cells isolated from preimplantation mouse embryos. *J. Embryol. exp. Morph.*, **45**: 37–53, 1978.

14. Hogan, B. and Tilly, R.: *In vitro* development of inner cell masses isolated immunosurgically from mouse blastocysts, I: Inner cell masses from 3.5 day p.c. blastocysts incubated for 24 h before immunosurgery. *J. Embryol. exp. Morph.*, **45**: 93–105, 1978.

15. Hogan, B. and Tilly, R.: *In vitro* development of inner cell masses isolated immunosurgically from mouse blastocysts, II: Inner cell masses from 3.5–4.0-day p.c. blastocysts. *J. Embryol. exp. Morph.*, **45**: 107–121, 1978.

16. Spindle, A. I.: Trophoblast regeneration by inner cell masses isolated from cultured mouse embryos. *J. Exp. Zool.*, **203**: 483–489, 1978.

17. Rossant, J. and Lis, W. T.: The possible dual origin of the ectoderm of the chorion in the mouse embryo. *Devel. Biol.*, **70**: 249–254, 1979.

18. Rossant, J. and Lis, W. T.: Potential of isolated mouse inner cell masses to form trophectoderm derivatives *in vivo*. *Devel. Biol.*, **70**: 255–261, 1979.

19. Gardner, R. L. and Johnson, M. H.: Investigation of cellular interaction and deployment in the early mammalian embryo using interspecific chimaeras between rat and mouse. In: Cell Patterning. Ciba Foundation Symposium 29 (new series), pp. 183–200. Elsevier/Excerpta Medica, Amsterdam, 1975.

20. Gardner, R. L., Papaioannou, V. E. and Barton, S. C.: Origin of the ectoplacental cone and secondary giant cells in mouse blastocysts reconstituted from isolated trophoblast and inner cell mass. *J. Embryol. exp. Morph.*, **30**: 561–572, 1973.

21. Rossant, J., Gardner, R. L. and Alexandre, H.: Investigation of the potency of cells from the postimplantation mouse embryo by blastocyst injection: a preliminary report. *J. Embryol. exp. Morph.*, **48**: 239–247, 1978.

22. Gardner, R. L.: The relationship between cell lineage and differentiation in the early mouse embryo. In: Results and Problems in Cell Differentiation, Vol. 9 (ed. W. J. Gehring), pp. 205–241. Springer-Verlag, Berlin, 1978.

23. Snell, G. D. and Stevens, L. C.: Early embryology. In: Biology of the Laboratory Mouse (ed. E. L. Green), pp. 205–245. McGraw-Hill, New York, 1966.

24. Rugh, R.: The Mouse. Burgess, Minneapolis, Minnesota, 1968.

25. Jolly, J. and Ferester-Tadie, M.: Recherches sur l'oeuf du rat et de la souris. *Arch. d'Anat. Microsc.*, **32**: 322–390, 1936.

26. Batten, B. E. and Haar, J. L.: Fine structural differentiation of the germ layers in the mouse at the time of mesoderm formation. *Anat. Rec.*, **194**: 125–142, 1979.

27. Solter, D., Damjanov, I. and Skreb, N.: Ultrastructure of the mouse egg cylinder. *Z. Anat. Entwickl.-Gesch.*, **132**: 191–199, 1970.

28. Gardner, R. L.: Manipulations on the blastocyst. *Adv. Biosc.*, **6**: 279–296, 1971.

29. Gardner, R. L. and Rossant, J.: Determination during embryogenesis. In:

Embryogenesis in Mammals. Ciba Foundation Symposium 40 (new series), pp. 5–18. Elsevier/Excerpta Medica, Amsterdam, 1976.

30. Gardner, R. L.: In preparation.
31. Rossant, J.: Cell commitment in early rodent development. In: Development in Mammals, Vol. 2 (ed. M. H. Johnson), pp. 119–150. Elsevier/North-Holland, Amsterdam, 1977.
32. Moustafa, L. A. and Brinster, R. L.: The fate of transplanted cells in mouse blastocysts *in vitro*. *J. exp. Zool.*, **181**: 181–192, 1972.
33. Moustafa, L. A. and Brinster, R. L.: Induced chimaerism by transplanting embryonic cells into mouse blastocysts. *J. exp. Zool.*, **181**: 193–202, 1972.
34. Grobstein, C.: Intra-ocular growth and differentiation of clusters of mouse embryonic shields cultured with and without primitive endoderm and in the presence of possible inductors. *J. exp. Zool.*, **119**: 355–380, 1952.
35. Kleinsmith, L. J. and Pierce, G. B.: Multipotentiality of single embryonal carcinoma cells. *Cancer Res.*, **24**: 1544–1551, 1964.
36. Levak-Svajger, B. and Svajger, A.: Differentiation of endodermal tissue in homografts of primitive ectoderm from two-layered rat embryonic shields. *Experientia*, **27**: 683–684, 1971.
37. Diwan, S. B. and Stevens, L. C.: Development of teratomas from the ectoderm of mouse egg cylinders. *J. Natl. Cancer Inst.*, **57**: 937–939, 1976.
38. Stevens, L. C. and Pierce, G. B.: Teratomas: definitions and terminology. In: Teratomas and Differentiation (ed. M. I. Sherman and D. Solter), pp. 13–14. Academic Press, London, 1975.
39. Skreb, N., Svajger, A. and Levak-Svajger, B.: Developmental potentialities of the germ layers in mammals. In: Embryogenesis in Mammals. Ciba Foundation Symposium 40 (new series), pp. 27–39. Elsevier/Excerpta Medica, Amsterdam, 1976.
40. Levak-Svajger, B. and Svajger, A.: Investigation on the origin of definitive endoderm in the rat embryo. *J. Embryol. exp. Morph.*, **32**: 445–459, 1974.
41. Svajger, A. and Levak-Svajger, B.: Regional developmental capacities of the rat embryonic ectoderm at the headfold stage. *J. Embryol. exp. Morph.*, **32**: 461–467, 1974.
42. Damjanov, I., Solter, D. & Skreb, N.: Teratocarcinogenesis as related to the age of embryos grafted under the kidney capsule. *Wilhelm Roux' Arch. Entwickl.-Mech. Org.*, **173**: 228–234, 1973.
43. Stevens, L. C.: The biology of teratomas. *Adv. Morphogen.*, **6**: 1–31, 1967.
44. Damjanov, I. and Solter, D.: Experimental teratoma. *Curr. Topics Path.*, **59**: 69–129, 1974.
45. Graham, C. F.: Teratocarcinoma cells and normal mouse embryogenesis. In: Concepts in Mammalian Embryogenesis (ed. M. I. Sherman), pp. 315–394. The MIT Press, Cambridge, Massachusetts, 1977.
46. Mintz, B. and Illmensee, K.: Normal genetically mosaic mice produced from malignant teratocarcinoma cells. *Proc. Nat. Acad. Sci. U.S.A.*, **72**: 3585–3589, 1975.

47. Papaioannou, V. E., McBurney, M. W., Gardner, R. L. and Evans, M. J.: Fate of teratocarcinoma cells injected into early mouse embryos. *Nature* (Lond.), **258**: 70–73, 1975.
48. Stevens, L. C. and Little, C. C.: Spontaneous testicular tumours in an inbred strain of mouse. *Proc. Nat. Acad. Sci. U.S.A.*, **40**: 1080–1087, 1954.
49. Stevens, L. C.: Experimental production of testicular teratomas in mice. *Proc. Nat. Acad. Sci. U.S.A.*, **52**: 654–661, 1964.
50. Stevens, L. C. and Varnum, D. S.: The development of teratomas from parthenogenetically activated ovarian mouse eggs. *Devel. Biol.*, **37**: 369–380, 1974.
51. Ozdenski, W.: Observations on the origins of primordial germ cells in the mouse. *Zool. Pol.*, **17**: 367–379, 1967.
52. Mintz, B., Cronmiller, C. and Custer, R. C.: Somatic cell origin of teratocarcinomas. *Proc. Nat. Acad. Sci. U.S.A.*, **75**: 2834–2838, 1978.
53. Mintz, B. and Russell, E. S.: Gene induced embryological modifications of primordial germ cells in the mouse. *J. exp. Zool.*, **134**: 207–239, 1957.
54. Stevens, L. C.: Experimental production of testicular teratomas in mice of strain 129, A/He and their F_1 hybrids. *J. natn. Cancer Inst.*, **44**: 929–932, 1970.
55. Papaioannou, V. E., Rossant, J. and Gardner, R. L.: Stem cells in early mammalian development. In: Stem Cells and Tissue Homeostasis (ed. B. I. Lord, C. S. Potten and R. J. Cole), pp. 49–69. Cambridge University Press, London, 1978.
56. Gartler, S. M. and Andina, R. J.: Mammalian X-chromosome inactivation. *Adv. Hum. Gen.*, **7**: 99–140, 1976.
57. Skreb, N. and Svajger, A.: Histogenic capacity of rat and mouse embryonic shields cultivated *in vitro*. *Wilhelm Roux' Arch. Entwickl.-Mech. Org.*, **173**: 228–234, 1973.
58. Skreb, N., Sculaneco-Spolifar, M. and Crnek, V.: Differentiation of teratomas during their development *in vitro*. *Bull. Sci. A.*, **21**: 137–138, 1976.
59. Skreb, N. and Crnek, V.: Tissue differentiation in ectopic grafts after cultivation of rat embryonic shields *in vitro*. *J. Embryol. exp. Morph.*, **42**: 127–134, 1977.
60. Skreb, N. and Svajger, A.: Experimental teratomas in rats. In: Teratomas and Differentiation (ed. M. I. Sherman and D. Solter), pp. 83–97. Academic Press, London, 1975.
61. Gardner, R. L.: Developmental potency of normal and neoplastic cells of the early mouse embryo. In: Birth Defects, Excerpta Medica International Congress Series No. 432 (ed. J. W. Littlefield and J. de Grouchy), pp. 154–166. Excerpta Medica, Amsterdam, 1974.
62. Whittingham, D. G.: Fertilisation, early development and storage of mammalian ova *in vitro*. In: The Early Development of Mammals (2nd Symposium of the British Society of Developmental Biology) (ed. M. Balls and A. E. Wild), pp. 1–24. Cambridge University Press, London, 1975.

63. Biggers, J. D. and Borland, R. M.: Physiological aspects of growth and development of the preimplantation mammalian embryo. *Ann. Rev. Physiol.*, **38**: 95–119, 1976.
64. Hsu, Y. C., Baskar, J., Stevens, L. C. and Rash, J. E.: Development *in vitro* of mouse embryos from the two cell stage to the early somite stage. *J. Embryol. exp. Morph.*, **31**: 235–245, 1974.
65. McLaren, A. and Hensleigh, H. C.: Culture of mammalian embryos over the implantation period. In: The Early Development of Mammals (2nd Symposium of the British Society for Developmental Biology) (ed. M. Balls and A. E. Wild), pp. 45–60. Cambridge University Press, London, 1975.
66. Wiley, L. M. and Pedersen, R. A.: Morphology of mouse egg cylinder development *in vitro*: A light and electron microscopic study. *J. exp. Zool.*, **200**: 389–402, 1977.
67. Hsu, Y. C.: *In vitro* development of individually cultured whole mouse embryos from blastocyst to early somite stage. *Devel. Biol.*, **68**: 453–461, 1979.
68. New, D. A. T.: Whole embryo culture and the study of mammalian embryos during organogenesis. *Biol. Rev.*, **53**: 81–122, 1978.
69. Buckley, S. K. L., Steele, C. E. and New, D. A. T.: *In vitro* development of early postimplantation rat embryos. *Devel. Biol.*, **65**: 396–403, 1978.
70. New, D. A. T., Coppola, P. T. and Terry, S.: Culture of explanted rat embryos in rotator tubes. *J. Reprod. Fertil.*, **35**: 135–138, 1973.
71. Beddington, R. S. P.: An autoradiographic analysis of the potency of embryonic ectoderm in the 8th day postimplantation mouse embryo. *J. Embryol. exp. Morph.*, **64**: 87–104, 1981.
72. Sadler, T. W.: Culture of early somite mouse embryos during organogenesis. *J. Embryol. exp. Morph.*, **49**: 17–25, 1979.
73. Tam, P. P. L. and Snow, M. H. L.: The *in vitro* culture of primitive-streak-stage mouse embryos. *J. Embryol. exp. Morph.*, **59**: 131–143, 1980.
74. Beddington, R. S. P.: An autoradiographic analysis of tissue potency in different regions of the embryonic ectoderm during gastrulation in the mouse. *J. Embryol. exp. Morph.* In press, 1982.
75. Balinsky, B. I.: An Introduction to Embryology, 4th Ed. W. B. Saunders Co., London, 1975.
76. Davidson, E. H.: Gene Activity in Early Development, 2nd Ed. Academic Press, Inc., New York, 1976.
77. Rosenquist, G. C.: A radioautographic study of labelled grafts in the chick blastoderm: Development from primitive streak stages to stage 12. *Contr. Embryol. Carneg. Instn.*, **38**: 71–110, 1966.
78. Nicolet, G.: Avian gastrulation. *Adv. Morphcgen.*, **9**: 231–262, 1971.
79. Skreb, N. and Frank, Z.: Developmental abnormalities in the rat induced by heat shock. *J. Embryol. exp. Morph.*, **11**: 445–457, 1963.
80. Skreb, N. and Bijelic, N.: Effects of X-rays on rat embryos during mesoderm formation. *Nature* (Lond.), **193**: 292–293, 1962.
81. Russell, L. B. and Russell, W. L.: An analysis of the changing radiation

response of the developing mouse embryo. *J. Cell. comp. Physiol.*, **43, Suppl. 1**: 103–147, 1954.

82. Snow, M. H. L. and Tam, P. P. L.: Is compensatory growth a complicating factor in mouse teratology? *Nature* (Lond.), **279**: 554–557, 1979.

83. Gehring, W. J.: Genetic control of determination in the *Drosophila* embryo. In: Genetic Mechanisms in Development. 31st Symposium of the Society for Developmental Biology (ed. F. Ruddle), pp. 103–128. Academic Press, Inc., New York, 1973.

84. Bennett, D.: *T*-locus mutants: suggestions for the control of early embryonic organisation through cell surface components. In: The Early Development of Mammals (2nd Symposium of the British Society for Developmental Biology) (ed. M. Balls and A. E. Wild), pp. 201–218. Cambridge University Press, London, 1975.

85. Bennett, D. and Dunn, L. C.: A lethal mutant (t^{w18}) in the house mouse showing partial duplications. *J. exp. Zool.*, **143**: 203–219, 1960.

86. Moser, G. C. and Gluecksohn-Waelsh, S.: Developmental genetics of a recessive allele at the complex *T*-locus in the mouse. *Devel. Biol.*, **16**: 564–576, 1967.

87. Artzt, K. and Bennett, D.: A genetically caused embryonal ectodermal tumour in the mouse. *J. natn. Cancer Inst.*, **48**: 141–158, 1972.

88. Bennett, D., Artzt, K., Magnusson, T. and Spiegelman, M.: Developmental interactions studied with experimental teratomas derived from mutants at the *T/t* locus in the mouse. In: Cell Interactions in Differentiation (ed. M. Karkinen-Jaaskelainen, L. Saxen and L. Weiss), pp. 389–398. Academic Press, Inc., New York, 1977.

89. Snow, M. H. L. and Bennett, D.: Gastrulation in the mouse: establishment of cell populations in the epiblast of t^{w18}/t^{w18} embryos. *J. Embryol. exp. Morph.*, **47**: 39–52, 1978.

90. Snow, M. H. L : Gastrulation in the mouse: growth and regionalisation of the epiblast. *J. Embryol. exp. Morph.*, **42**: 293–303, 1977.

91. Snow, M. H. L.: Proliferative centres in embryonic development. In: Development in Mammals, Vol. 3 (ed. M. H. Johnson), pp. 337–362. Elsevier/North-Holland, Amsterdam, 1978.

92. Pasteels, J.: Etudes sur la gastrulation des vertebres meroblastiques, III: Oiseaux; IV: Conclusions generales. *Archiv. Biol.*, **48**: 381–488, 1937.

93. Wiley, L. M., Spindle, A. I. and Pedersen, R. A.: Morphology of isolated mouse inner cell masses developing *in vivo*. *Devel. Biol.*, **63**: 1–10, 1978.

Discussion

Dr. Le Douarin: How long is it possible to culture postimplantation embryos? If the injected cells fall inside the amniotic cavity, can they be incorporated inside the embryo thereafter, or are they lost?

Dr. Beddington: The length of time over which postimplantation embryos will develop normally *in vitro* depends on both the species and the stage of explantation. Pre-primitive streak stage rat embryos can develop normally for over 4 days, reaching the 30–40 somite stage.[69] No comparable success has been achieved with mouse embryos. After 48 hours in culture pre-primitive streak stage mouse embryos show a reduction in total protein content compared with embryos of an equivalent gestational age which have developed *in utero*.[73] However, late primitive streak stage mouse embryos will develop normally for 36 hours,[71] and embryos explanted at the early somite stage develop in parallel with embryos maintained *in vivo* for at least 2 days.[72] The superiority of rat embryo culture is probably due to the fact that more time has been spent working out the postimplantation culture conditions for this species than for other mammalian embryos.[68]

The fate of injected cells which fall inside the amniotic cavity is not known for certain, but it would seem unlikely that free floating cells could incorporate into the epithelial layer of the embryonic ectoderm. Many *in vitro* chimeras contain a few very densely labeled cells either floating in the amniotic cavity or adhering to the amnion. These presumably represent those injected cells which failed to incorporate into the embryo at the time of injection and subsequently died. Some injected embryos fail to show any trace of chimerism on autoradiographic analysis, and therefore it is likely that donor cells can be extruded from the embryo and lost in the culture medium.

Dr. Stevens: What kinds of instruments do you use?

Dr. Beddington: Glass needles.

Dr. Mullen: Is it known for sure that the embryonic endoderm is derived from inner cells in the morula or early blastocyst? Could they be migrating in from the trophectoderm or outer cells in the morula?

Dr. Beddington: It is clear from blastocyst injections that ICM cells from the 4th day blastocyst can give rise to primary endoderm, thereby generating chimerism in the visceral yolk sac endoderm. In addition, ICMs isolated by microsurgery from 4th day blastocysts[10] and ICMs recovered immunosurgically from 4th day blastocysts and grown in culture for 24 hours[14, 93] form two-layered structures consisting of an inner core of embryonic ectoderm surrounded by a layer of primary endoderm. Trophectoderm vesicles, on the other hand, have never been reported to give rise to primary endoderm.[8] The obvious conclusion must be that primary endoderm is derived from the ICM rather than the trophectoderm.

Dr. Okada: When smooth (embryonic ectoderm) and rough (primary endoderm) cells are mixed in cell culture conditions, do they selectively adhere?

Dr. Beddington: This experiment has never been done, partly because isolated embryonic ectoderm cells do not survive in culture. If disaggregation is incomplete then an isolated rind of primary endoderm will wrap around a core of embryonic ectoderm (R. L. Gardner, personal communication).

Dr. Le Douarin: In the chick embryo the primitive endoderm gives rise to the anterior and lateral areas of the yolk sac endoderm whilst the posterior yolk sac endoderm arises from the ectoblast. Is it the same picture in the mouse?

Dr. Beddington: The origin of different regions of the yolk sac endoderm has not been examined directly in the mouse. However, in blastocyst injection chimeras, embryonic ectoderm from the 5th day blastocyst does not contribute to the yolk sac endoderm, and colonization of this tissue has never been observed in *in vitro* chimeras injected with 8th day embryonic ectoderm. This indicates that all the yolk sac endoderm in the mouse is derived from the primary endoderm.

Dr. Yutaka Toyoda: Would you please comment on the cell lineage in the germ line, that is, the origin of primordial germ cells?

Dr. Beddington: It is known that the embryonic ectoderm from early post-implantation embryos can give rise to teratocarcinomas[37] and therefore to cells, EC cells, which can be considered totipotent.[46, 47] That embryonic ectoderm can generate totipotent cells in ectopic grafts sets a precedent for this tissue's also being the precursor of the primordial germ cells.

Dr. Tachi: I think the possibility that Dr. Mullen has suggested is very likely. But actual analysis of the patterns suggests a very complex event underlying the formation of the patterns in the chimeras. We are trying to obtain clues to approach this problem by developing a technique to analyze the patterns created in chimeric mice, possibly by means of a computer simulation.

Dr. Okada: It seems to me most critical to amalgamate the concept of "cell lineage" with "determination" in order to understand cell differentiation in development. When a cell differentiated in A still has the potential to switch into B and C, but not into D, E and others, can we simply assume that A, B and C belong to the same line in cell lineage? I doubt it.

Early Stages of Fertilization *in vitro* of Rabbit, Pig, Cattle and Human Eggs

Akira Iritani

The present manuscript reviews recent studies performed in our laboratory on fertilization *in vitro* in large animal species. Species differences in the capacitation of spermatozoa and the early process of fertilization will be also discussed.

I. EXPERIMENTS IN RABBITS[1]

The experimental method used was modified from that of Brackett and Oliphant (1975).[2] It has been established that rabbit epididymal spermatozoa can be capacitated completely during preincubation in a defined medium with successful fertilization *in vitro* of ovulated eggs.

Materials and methods

The composition of the defined medium used for manipulation of gametes is shown in Table 1. The theoretical osmolarity of the medium was 332.41. The epididymal spermatozoa were collected by flushing the caudal epididymal duct through the vas deferens with 2 ml medium, they were then suspended in 3–5 ml medium and washed by centrifugation at 350 *g* for 5 min. The sedimented spermatozoa were resuspended in 3 ml

TABLE 1. Composition of the Medium Used for Fertilization *in vitro* in Rabbits

NaCl	112.00 mM
KCl	4.02
CaCl$_2$	2.25
NaH$_2$PO$_4$	0.83
MgCl$_2$	0.52
NaHCO$_3$	37.00
Glucose	13.90
Sodium pyruvate	1.25
Bovine serum albumin	3 mg/ml
Potassium penicillin G	31 μg/ml

Department of Animal Science, College of Agriculture, Kyoto University, Kyoto, Japan

medium and incubated for 15 min in a water bath at 37° C in air and then centrifuged again (the second washing). The spermatozoa were suspended again in 0.5–1.0 ml medium (0.25–1.0 × 10^8 spermatozoa/ml). A portion (0.02 ml) of the sperm suspension was introduced into 1 ml medium in a plastic culture dish. The diluted sperm suspension (0.5–2.0 × 10^6 spermatozoa/ml) was preincubated for 10–10.5 h in a CO_2 incubator at 37° C. The mature does were induced to superovulate by injections of FSH followed by hCG. The eggs were flushed from the oviducts of the does 14–15 h after the injection of hCG. The eggs were washed twice with the medium, and then introduced into the sperm suspension. After the eggs were incubated with the spermatozoa for 1–5.25 h in a CO_2 incubator, they were fixed with glutaraldehyde and 10% neutral formalin and stained with 0.25 % lacmoid.

Results and discussion

As shown in Table 2, the spermatozoa start to penetrate the eggs within 1–1.25 h after the introduction of eggs into sperm suspension. This

TABLE 2. Fertilization *in vitro* of Rabbit Eggs by Epididymal Spermatozoa Preincubated for 10–10.5 h after Treatment for 15 Min with the Medium[a]

Time of examination (h after incubation)	No. of eggs penetrated			Total no. eggs penetrated /no. eggs examined (%)
	Sperm in perivitellin space	With enlarged sperm head	With pronuclei	
1–1.25	1	5	0	6/18 (33)
2–2.25	2	10	0	12/20 (60)
3–3.25	2	12	4	18/23 (78)
4–4.25	4	5	13	22/27 (81)
5–5.25	6	5	17	28/38 (74)

[a] Percentage of motile spermatozoa at insemination was 50–60%.

indicates that the spermatazoa are completely capacitated during the preincubation period. Penetration into the eggs was almost complete within 2.25 and 3.25 h. The first appearance of the male pronucleus in the penetrated eggs was observed 3–3.25 h after insemination. The male and female pronuclei were still in early stage at 3–3.25 h after insemination, but they showed distinct nuclear membranes 2 h later. Most of the penetrated eggs developed to the 2–4-cell and 8–16-cell stages 24 and 48 h after insemination, respectively.

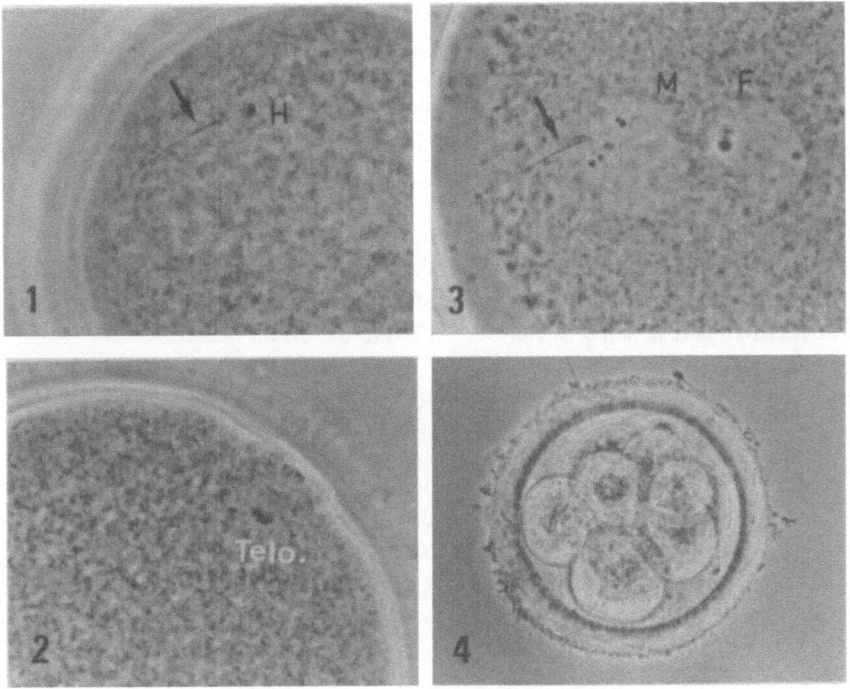

PLATE 1. Fertilization *in vitro* of rabbit eggs.[1]
FIG. 1. Fertilizing spermatozoon in an egg fixed 1 h after insemination. The detached sperm tail (arrow) and slightly swollen sperm head (H) are seen. ($\times$ 345)
FIG. 2. An egg fixed 1.25 h after insemination. The second meiotic division of telophase is seen. ($\times$ 360)
FIG. 3. Early pronuclear egg fixed 3 h after insemination. Male (M) and female (F) pronuclei, and the detached sperm tail (arrow) are seen. ($\times$ 380)
FIG. 4. An unstained egg at the 8–16-cell stage observed 48 h after insemination. ($\times$ 180)

It seems that pretreatment of spermatozoa for 15 min with the medium followed by the second washing is one of the most important procedures for completion of *in vitro* sperm capacitation in the rabbit.

II. EXPERIMENTS IN CATTLE[3]

The maturation of follicular oocytes in culture and fertilization after transfer to mated animals have been reported for pigs[4] and cattle.[5,6] However, there do not appear to be any reports on the fertilization *in vitro* of follicular oocytes matured *in vitro* for large domestic animals such as cattle. The present experiment was performed to examine the feasibility of capacitation of bovine spermatozoa in reproductive tracts isolated from estrous cows or in the uteri *in situ* of estrous rabbits.

Materials and methods

Composition of the medium used for manipulation of the gametes was the modified Kreb's Ringer bicarbonate solution (m-KRB) which was used for fertilization *in vitro* in the rat.[7] (Table 3)

TABLE 3.   Composition of Modified KRB Solution

NaCl	94.6 mM
KCl	4.78
CaCl$_2$	1.71
KH$_2$PO$_4$	1.19
MgSO$_4$	1.19
NaHCO$_3$	25.07
Sodium lactate	21.58
Sodium pyruvate	0.50
Glucose	5.56
Crystalline bovine serum albumin	4 mg/ml
Streptomycin sulphate	50 μg/ml
Potassium penicillin	75 μg/ml

Follicular oocytes were collected at 30° C by puncturing follicles with a needle from the ovaries randomly removed at a local slaughterhouse. The collected oocytes were washed twice with m-KRB solution. The washed oocytes were cultured for 20–24 h at 37° C in a CO$_2$ incubator until the second metaphase was reached.[8] A 3-ml aliquot of freshly collected semen was washed once with m-KRB solution, and the washed spermatozoa were resuspended in the medium. One of four portions of the washed spermatozoa was incubated for 12–14 h at 37° C in m-KRB solution in a CO$_2$ incubator. Two portions were incubated for 3–4 h in the isolated genital tract from an estrous cow. The volume of sperm suspension and the number of spermatozoa incubated were 0.3 ml and 1.5×10^8 in the uterine horn and 0.1 ml and 5×10^7 in the oviduct, respectively. The tract was kept in saline at 37° C. The fourth portion of the sperm suspension (0.1 ml, 5×10^7 spermatozoa) was introduced into the uterus of a doe pretreated with FSH followed 12 h later by hCG. The does were kept for 12–14 h before being killed for recovery of the spermatozoa. After incubation the spermatozoa were recovered by flushing the cow oviducts and uteri and the rabbit uteri. All sperm samples were washed by centrifugation at 500 g for 10 min. Eggs were fixed for 3–4 days in 25% acetic alcohol 18–21 h after insemination and stained with 1% aceto orcein, and then examined using phase-contrast microscopy.

Results and discussion

As shown in Table 4, 60.1% (141/232) of the follicular oocytes normally matured to the second metaphase. This figure was slightly lower than those obtained in our previous experiments in which 66.7% of oocytes matured normally under very similar conditions.[8] None of the 38 oocytes inseminated with spermatozoa preincubated in m-KRB solution appeared to be fertilized but about 20% showed clear signs of sperm penetration followed by activation to the pronuclear stage with insemination of spermatozoa preincubated in the reproductive tract of the estrous cow or in the rabbit uterus. The capacitation of bull spermatozoa in the uterus of an estrous rabbit suggests that the rabbit uterus could be used for the study of the capacitation requirements of spermatozoa from different species.

Since experiments with pig oocytes suggested that epididymal spermatozoa can be capacitated more easily than ejaculated spermatozoa, we investigated the feasibility of capacitation of bull epididymal spermatozoa in a chemically defined medium, m-KRB solution.

The frozen-stored epididymal spermatozoa were thawed at 40°C and washed once with m-KRB solution. The washed spermatozoa were resuspended in the medium (3×10^6 cells/0.5 ml), and then preincubated for 4 h at 37°C in a CO_2 incubator. The procedures for the maturation of follicular oocytes, insemination and examination of oocytes were the same as previously described.

TABLE 4. Fertilization *in viro* of Cattle Follicular Oocytes Matured in Culture

Preincubation of spermatozoa	Total no. of oocytes cultured	No. of oocytes maturing to Met.-II (%)	No. of oocytes fertilized (%)
12–14 h in m-KRB	62	38 (61)	0/38 (0)
3–4 h in cow oviduct or uterus	92	56 (61)	11/56 (20)
12–14 h in rabbit uterus	78	47 (60)	10/47 (21)

The results, summarized in Table 5, show that a comparatively high percentage of penetrated oocytes (5/19 = 26.3%) was obtained. This figure is similar to that obtained in the oocytes inseminated with ejaculated spermatozoa preincubated in the genital tract, indicating that epididymal spermatozoa can be effectively capacitated *in vitro*.

TABLE 5. Fertilization *in vitro* of Cattle Follicular Oocytes with Frozen-stored epididymal Spermatozoa Capacitated in a Defined Medium (Iritani *et al.*, unpublished)

| No. of trials | No. of oocytes cultured | No. of oocytes maturing to Met.-II (%) | No. of oocytes penetrated | | |
			Total (%) Telo.-II	Female pronucleus	Male pronucleus
2	35	19/35 (54)	5/19 (26) 2	2	1

III. EXPERIMENTS IN THE PIG[9]

Although Motlik and Fulka[4] reported the successful fertilization of pig follicular oocytes matured in culture and transfer to mated animals, we have been unable to find any reports on the fertilization *in vitro* of pig oocytes matured in culture. This study was conducted to examine the feasibility of capacitation of boar epididymal and ejaculated spermatozoa in reproductive tracts isolated from estrous sows.

Materials and methods
Procedures for the preparation of the medium, oocytes and spermatozoa, insemination and examination of eggs were almost the same as those used in the experiments in cattle. Ejaculated and epididymal spermatozoa were collected from the same boars before and after slaughter, and they were washed twice and preincubated for 4.5–5 h at 37° C as shown in Table 6. The volume of sperm suspension and the number of spermatozoa incubated were 0.1 ml (4×10^7) in m-KRB and in the oviduct, and 0.5 ml (2×10^8) in the uterine horn. After incubation, the spermatozoa were recovered by flushing the oviducts and uteri. Then the suspension was washed once and centrifuged. A 10-μl aliquot of the sperm suspension (0.5–1×10^6 spermatozoa/ml) was introduced into the 0.4 ml medium containing the 10–15 cultured oocytes. The oocytes were cultured for an additional 17–20 h at 37° C, and then they were fixed and stained for examination using phase-contrast microscopy.

TABLE 6. Fertilization *in vitro* of Pig Follicular Oocytes Matured in Culture

Preincubation of spermatozoa for 4.5–5 h	Total no. of oocytes cultured	No. of oocytes maturing to Met.-II (%)	No. of oocytes fertilized (%)
Ejaculated			
m-KRB	32	19 (59)	0/19 (0)
Oviduct or uterus	88	51 (58)	12/51 (24)
Epididymal			
m-KRB	30	17 (57)	1/17 (6)
Oviduct or uterus	99	64 (65)	23/64 (36)

Results and discussion

As shown in Table 6, 151 out of 249 (60.6%) oocytes cultured matured to the second metaphase. This figure was comparable to that of 60.9% which was obtained in our previous experiments.[10] Fertilization by ejaculated spermatozoa preincubated in m-KRB did not occur. Eggs incubated with spermatozoa preincubated in the oviduct or uterus showed clear signs of penetration and activation; in this case, the fertilization rate was 24% (12/51). When preincubated epididymal spermatozoa were used for insemination a higher percentage of fertilized eggs (23/64 = 36%) was obtained. Only one of the 16 eggs was penetrated by epididymal spermatozoa preincubated in m-KRB. The slightly higher proportion of oocytes penetrated by epididymal than ejaculated spermatozoa indicates that sperm-coating factors may be involved in the poorer capacitation of ejaculated spermatozoa, as suggested by Hunter, Holtz and Henfrey.[11] In the present study using pig oocytes matured in culture, various abnormalities, such as asynchronous development of the male and female pronuclei and delay or abnormal transformation of the sperm head into a pronucleus, were observed. However, some oocytes normally developed to the pronuclear stage (Pl. 2, Fig. 4).

IV. EXPERIMENTS IN HUMANS[12]

Although clear cytological evidence of *in vitro* fertilization of human oocytes matured in various media containing human follicular fluid has been reported,[13-15] successful maturation and sperm penetration *in vitro* has not yet appeared in the literature. In this experiment we sought to establish a system in which human follicular oocytes obtained from excised

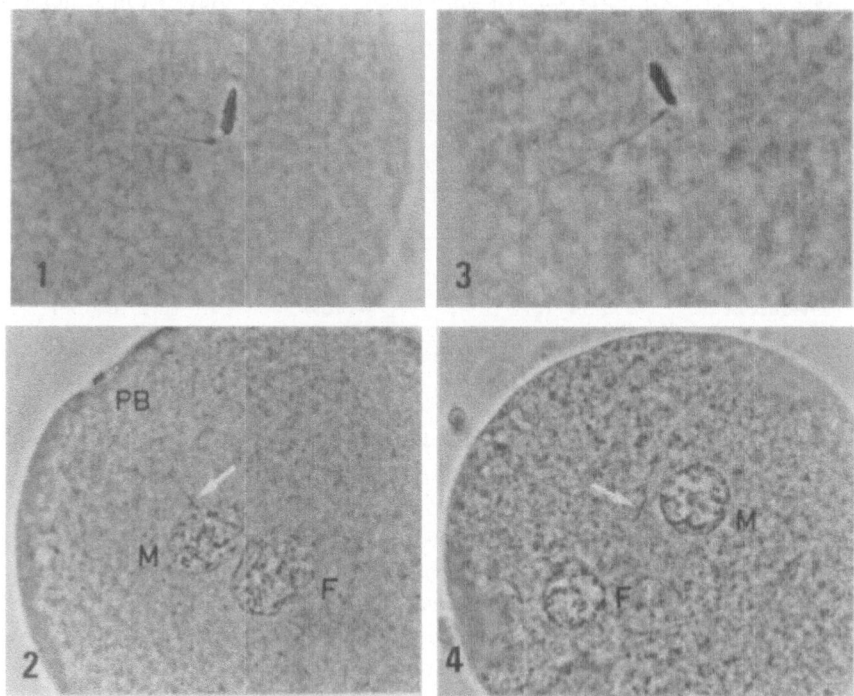

PLATE 2. Fertilization *in vitro* of pig and cattle eggs.[3]

FIG. 1. The fertilizing bull spermatozoon in an oocyte fixed 18 h after the introduction of spermatozoa preincubated in a rabbit uterus for 14 h. ($\times$ 400)

Fig. 2. A pronuclear cow egg fixed 18 h after the introduction of spermatozoa preincubated for 3.5 h in the isolated uterus from an estrous cow. The sperm mid-piece (arrow), male pronucleus (M), female pronucleus (F) and the second polar body (PB) are seen. ($\times$ 220)

FIG. 3. The fertilizing pig spermatozoon fixed 17 h after the introduction of epididymal spermatozoa preincubated in the isolated uterus from an estrous sow for 5 h. ($\times$ 540)

FIG. 4. A pronuclear pig egg fixed 20 h after the introduction of spermatozoa preincubated in the isolated uterus from an estrous sow. The sperm mid-piece (arrow), male pronucleus (M) and female pronucleus (F) are seen. ($\times$ 195)

ovaries could be matured in a defined medium with successful fertilization *in vitro* by capacitated spermatozoa.

Materials and methods

The medium used for the manipulation of gametes was the m-KRB solution used in previous experiments in pig and cattle oocytes. The oocytes were randomly collected by puncturing follicles within 2 h after removal of the ovaries. The collected oocytes were examined under a dissecting microscope, and those which had degenerated were not used for culture. Morphologically normal oocytes were washed twice with the medium and cultured for 40–40.5 h in a CO_2 incubator. Then the oocytes were again examined under a microscope, and those showing clear cytoplasmic degeneration were excluded. The remaining oocytes were washed twice and exposed to preincubated spermatozoa. Semen specimens were obtained from fertile donors, liquefied and washed twice with m-KRB solution. The diluted sperm suspension including 0.5–1.0×10^6 spermatozoa/ml was preincubated for 3 h in a CO_2 incubator. After the cultured oocytes were incubated with the spermatozoa for 10 h, the oocytes were fixed with 2.5% glutaraldehyde and 10% neutral formalin. They were then stained with 0.25% lacmoid in 45% acetic acid and examined for evidence of sperm penetration.

Results and discussion

As shown in Table 7, 17 of the 48 oocytes (35%) matured to the second metaphase after the additional culture for 10 h with spermatozoa. Four-

TABLE 7.   Fertilization *in vitro* of Human Follicular Oocytes Matured in Culture

No. of oocytes inseminated	No. of oocytes maturing to Met.-II at examination (%)	No. of oocytes fertilized (%)
48	17/48 (35)	14/17 (82)

teen of the 17 normally matured oocytes (82%) were fertilized with female pronucleus, male pronucleus, sperm mid-piece and the second polar body. On the other hand, 5 of the 16 oocytes which matured to the first metaphase were penetrated, but the sperm head remained swollen in the penetrated oocytes (Pl. 3, Fig. 1). The same observations were also re-

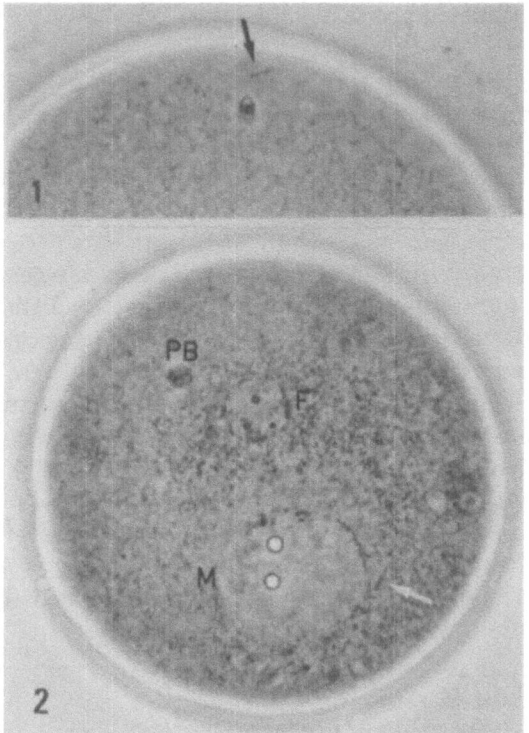

PLATE 3. Fertilization *in vitro* of human eggs.[12]
FIG. 1. The fertilizing spermatozoon in an oocyte fixed 10 h after insemination with spermatozoa preincubated for 3 h, showing the sperm mid-piece and slightly swollen sperm head. ($\times$ 560)
FIG. 2. A pronuclear egg fixed 10 h after insemination with spermatozoa preincubated for 3 h. The sperm mid-piece (arrow), male pronucleus (M), female pronucleus (F) and the second polar body (PB) are seen. ($\times$ 360)

ported for mouse[16] and rat[17] oocytes. Thus, the maturation of ooplasm may be of importance for normal fertilization.

Acknowledgements

This work was supported by a grant from the Ford Foundation (No. 740-0404) and from the Ministry of Education of Japan (No. 448067).

REFERENCES

1. Hosoi, Y., Niwa, K., Hatanaka, S. and Iritani, A.: Fertilization *in vitro* of rabbit eggs by epididymal spermatozoa capacitated in a chemically defined medium. *Biol. Reprod.*, **24**: 637–642, 1981.
2. Brackett, B. G. and Oliphant, G.: Capacitation of rabbit spermatozoa *in vitro*. *Biol. Reprod.*, **12**: 260–274, 1975.
3. Iritani, A. and Niwa, K.: Capacitation of bull spermatozoa and fertilization *in vitro* of cattle follicular oocytes matured in culture. *J. Reprod. Fert.*, **50**: 119–121, 1977.
4. Motlik, J. and Fulka, J.: Fertilization of pig follicular oocytes cultivated *in vitro*. *J. Reprod. Fert.*, **26**: 387–389, 1974.
5. Sreenan, J.: *In vitro* maturation and attempted fertilization of cattle follicular oocytes. *J. agric. Sci. Camb.*, **75**: 393–396, 1970.
6. Hunter, R. H. F., Lawson, R. A. S. and Rowson, L. E. A.: Maturation, transplantation and fertilization of ovarian oocytes in cattle. *J. Reprod. Fert.*, **30**: 325–328, 1972.
7. Toyoda, Y. and Chang, M. C.: Fertilization of rat eggs *in vitro* by epididymal spermatozoa and the development of such eggs following transfer. *J. Reprod. Fert.*, **36**: 9–22, 1974.
8. Sato, E., Iritani, A. and Nishikawa, Y.: Factors involved in maturation of pig and cattle follicular oocytes cultured *in vitro*. *Jap. J. Anim. Reprod.*, **23**: 12–18, 1977.
9. Iritani, A., Niwa, K. and Imai, H.: Sperm penetration *in vitro* of pig follicular oocytes matured in culture. *J. Reprod. Fert.*, **54**: 379–383, 1978.
10. Sato, E., Iritani, A. and Nishikawa, Y.: Rate of maturation division of pig follicular oocytes. *Jap. J. Zootech. Sci.*, **49**: 400–405, 1978.
11. Hunter, R. H. F., Holtz, W. and Henfrey, P. J.: Epididymal function in the boar in relation to the fertilizing ability of spermatozoa. *J. Reprod. Fert.*, **46**: 463–466, 1976.
12. Nishimoto, T., Yamada, I., Niwa, K., Mori, T., Nishimura, T. and Iritani, A.: Sperm penetration *in vitro* of human oocytes matured in a chemically defined medium. *J. Reprod. Fert.,* **64**: 115–119, 1982.
13. Edwards, R. G., Bavister, B. D. and Steptoe, P. C.: Early stages of fertilization *in vitro* of human oocytes matured *in vitro*. *Nature* (Lond.), **221**: 632–635, 1969.
14. Bavister, B. D., Edwards, R. G. and Steptoe, P. C.: Identification of the midpiece and tail of the spermatozoa during fertilization of human eggs *in vitro*. *J. Reprod. Fert.*, **20**: 159–160, 1969.
15. Soupart, P. and Morgenstern, L. L.: Human sperm capacitation and *in vitro* fertilization. *Fert. Steril.*, **24**: 462–478., 1973.
16. Iwamatsu T. and Chang, M. C.: Sperm penetration *in vitro* of mouse oocytes at various times during maturation. *J. Reprod. Fert.*, **31**: 237–247, 1972.

17. Niwa, K. and Chang, M. C.: Fertilization of rat eggs *in vitro* at various times after ovulation with special reference to fertilization of ovarian oocytes matured in culture. *J. Repord. Fert.*, **43**: 435–451, 1975.

Discussion

Dr. Kato: It has been made clear that fertilized oocytes *in vitro* do not cleave and develop to advanced stages in culture in some mammalian species such as hamsters and rats. What do you think about this?

Dr. Iritani: Yes, I know that a hamster egg fertilized *in vitro* and/or *in vivo* does not cleave to 4-, 8- or 16-cell stages in culture. I do not know why they do not cleave, but I think some necessary factors included in the oviductal environments may be absent in the *in vitro* system. It is also suggested that hamster and rat fertilized eggs are more sensitive to the environments different from the oviducts.

Dr. Tone: You showed that the fertilized eggs could not yet mature in human *in vitro* fertilization experiments. What is the relationship between egg maturation and sperm penetration or fertilization?

Dr. Iritani: We have used different terms. Sperm penetration means that the capacitated spermatozoon just penetrated an oocyte and remained slightly swollen, but did not transform into a pronucleus while the penetrated oocyte was not fully matured. When a spermatozoon penetrated a matured oocyte (Metaphase II), the sperm head could normally be transformed into the pronucleus. This fact was confirmed in the mouse, rat, pig and human oocyte.

Early Development of Mouse Embryos Fertilized *in vitro*

Yutaka Toyoda and Makoto Takasugi

INTRODUCTION

A major advantage of *in vitro* fertilization as a research tool is that the start of fertilization can be controlled more precisely than with *in vivo* studies by eliminating numerous factors involved in the approximation of gametes in the female reproductive tract. This would certainly be useful. for all investigators concerned with the genetic aspects of developmental neurobiology. Although *in vitro* studies of the preimplantation stages of the mouse have been reported[1,2] and the fertilization *in vitro* of mouse eggs has been observed by several investigators,[3,4] detailed aspects of preimplantation development have not been fully explored. In this paper, the characteristics of the mouse embryo fertilized *in vitro* will be discussed with reference to their development in culture.

MATERIALS AND METHODS

In the present study, epididymal spermatozoa were obtained from mature male mice, suspended in a medium under paraffin oil, and pre-incubated for 1–2 h at 37° C under 5 % CO_2 in air.[5] Superovulated oocytes were collected from excised oviducts of mature female mice which had received i.p. injections of 5 IU PMSG and 5 IU hCG 48 h apart. The females were killed 15–16 h after hCG injection. A small volume of prein-cubated sperm suspension was introduced to the medium containing the oocytes with cumulus cells. The mixture of eggs and spermatozoa was then incubated at 37° C under 5 % CO_2 in air. Sperm concentration in the fertilization medium was adjusted to 100–150 cells/μl in each experiment. The inseminated oocytes were transferred to fresh medium 6 h after insemination and cultured further for 90 h. Embryos were examined for the

School of Veterinary Medicine and Animal Sciences, Kitasato University, Towada-shi, Aomori, Japan first cleavage at one hour intervals between 16 and 24 h after insemination, then the inspection was continued at 12 h intervals.

The chemical compositions of the culture media used are shown in Table 1. Fertilization in TYH medium[6] is highly reproducible, demonstrating

TABLE 1. Chemical Composition of Defined Media Used for *in vitro* Fertilization and Culture of Mouse Embryos

Compound	Concentration (mM)	
	TYH	TYH-280
NaCl	110.37	72.13
KCl	4.78	4.78
$CaCl_2 \cdot 2H_2O$	1.71	—
Ca lactate	—	1.71
KH_2PO_4	1.19	1.19
$MgSO_4 \cdot 7H_2O$	1.19	0.85
$NaHCO_3$	25.07	25.07
Na lactate	—	30.00
Na pyruvate	1.00	0.30
Glucose	5.56	5.56
B.S.A.	4 mg/ml	4 mg/ml
Penicillin	100 U/ml	100 U/ml
Streptomycin	50 μg/ml	50 μg/ml
Reference	6	7

that sodium lactate is not essential for sperm capacitation and fertilization in the mouse. TYH-280 medium was modified by Kasai *et al.*[7] for culturing mouse embryos fertilized *in vitro*. The modification is based on the findings of Cross and Brinster[8] which demonstrated separate energy requirements for the first cleavage division versus development thereafter. In this experiment, embryos were cultured either in TYH or TYH-280 medium under paraffin oil at 37°C under 5% CO_2 in air, although lower oxygen concentrations have been reported to be more favorable for preimplantation development.[2,9]

RESULTS AND DISCUSSION

Sperm penetration and the first cleavage

Using scanning electron microscopy, Motomura and Toyoda[10] found that the spermatozoa passed the cumulus oophorus, reached the surface of the zona pellucida, and began to penetrate it within 5 min of insemination. The penetration was complete in most oocytes after 20 min. Resumption of the second meiotic division was observed by light microscopy in about 80% of the penetrated eggs at 45 min after insemination, anaphase II being predominant at this stage. Most eggs reached telophase II by 60 min after insemination.[5]

Based on these findings, it is possible to obtain eggs at a specific meiotic stage in order to study the early events following fertilization. More accurate control of sperm penetration might be accomplished by improving techniques and procedures for sperm preincubation and fertilization. Accelerated sperm penetration was achieved by a brief preincubation of spermatozoa in a medium with high ionic strength[11] and in the presence of caffeine.[12] It has also been suggested by Fraser and Quinn[13] and Okamoto and Toyoda[14] that sperm penetration of mouse eggs *in vitro* may be triggered by the addition of glucose to glucose deficient medium which allows the gametes to survive but does not allow fusion.

The timing of the first cleavage division of mouse embryos fertillized *in vitro* are shown in Fig. 1. Two kinds of eggs, i.e., ICR and C3H/He ×

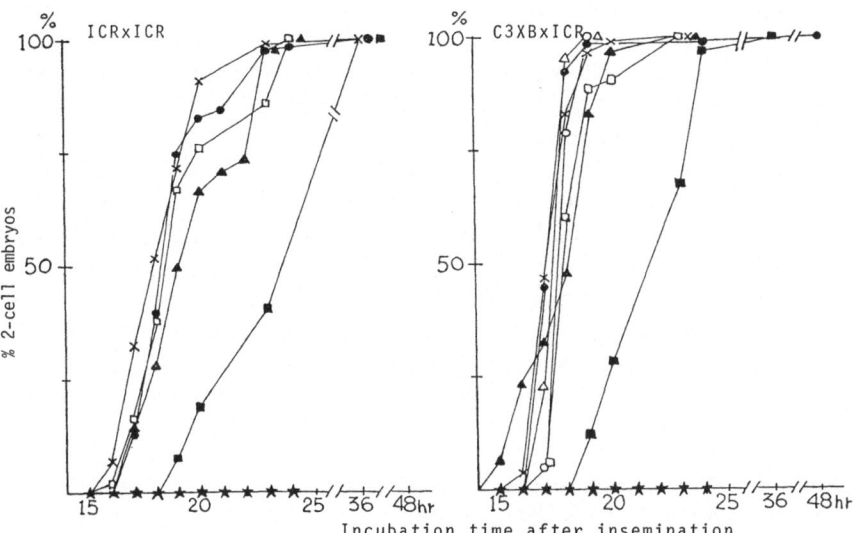

FIG. 1. Cumulative percentage of 2-cell embryos of ICR (left) and C3 × B (right) eggs fertilized by ICR spermatozoa. The eggs were cultured in TYH media containing different concentrations of sodium pyruvate. Concentrations were 8 mM (○), 4 mM, (△), 2 mM (□), 1 mM (×), 0.5 mM (●), 0.25 mM (▲), 0.125 mM (■) and 0 mM (★).

C57BL/6J-at F_1 (C3 × B) eggs, were fertilized with ICR spermatozoa and cultured in TYH medium containing various concentrations of sodium pyruvate. The first cleavage is apparently dependent on the presence of pyruvate in the medium, as reported by Biggers *et al.*[15] for one-cell mouse zygotes fertilized *in vivo*. In this study, the first cleavage was retarded when the level of pyruvate was reduced to 0.125 mM and all of the eggs degenerated in the absence of pyruvate. Median time of the first cleavage was 17–18 h after insemination in the presence of adequate amount of pyruvate, in agreement with the time reported by Kaufman.[16] Practically all of the eggs developed to the 2-cell stage within 24 h of incubation. Strain differences in the time required for the completion of fertilization have been reported for the mouse embryos fertilized *in vitro*[17] as well as for those fertilized *in vivo*.[18,19] but in the latter case, the maternal and paternal effects seen in *in vivo* fertilization may involve the approximation of gametes, especially in the transport of spermatozoa in the female reproductive tract.

Development beyond the 2-cell stage

In marked contrast to the events leading to the first cleavage, development beyond the 2-cell stage was quite different in the two types of eggs. As shown in Table 2, almost all 2-cell embryos (145/147) derived from C3 × B eggs developed to the 4-cell stage in 24 h and 93% of the cultured embryos successfully developed to blastocysts in 72 h in TYH-280 medium. On the other hand, only 16% of the 2-cell embryos derived from ICR eggs fertilized by ICR spermatozoa *in vitro* reached the 4-cell stage in 24 h and a still smaller proportion (3.2%) of the embryos finally developed to the blastocyst stage. Thus, the second cleavage is critical for *in vitro* fertilized mouse zygotes to proceed to the advanced stages of development. The ability to surmount this critical stage seems to be largely dependent on the maternal genetic effect rather than the embryonic genome itself.

TABLE 2. Development beyond the 2-cell Stage of *in vitro* Fertilized Mouse Embryos Cultured in TYH-280 Medium

Strain		No. of 2-cell embryos cultured	No. (%) of embryos developed to:	
Egg	Sperm		4–8 cell stage in 24 h	blastocyst in 72 h
ICR	ICR	126	20 (15.9)	4 (3.2)
ICR	C3×B*	141	48 (34.0)	22 (15.6)
C3×B*	ICR	147	145 (98.6)	137 (93.2)

*C3×B: C3H/He × C57BL/6J-at

Essentially the same results have been obtained by Whitten and Biggers[1] for the development of one-cell mouse embryos fertilized *in vivo*. They have shown that one-cell embryos derived from C57BL/10J × SJL/J F_1 females successfully developed to blastocysts, while only a few embryos derived from either inbred females could do so. The cause of this difficulty in culturing embryos beyond the 2-cell stage is not fully understood, but it may be related to the long G2 phase between the 2-cell and 4-cell stage,[20] which has been estimated to last for about 12 h[21] or for 15–18 h.[22] Abramczuk *et al.*[23] obtained successful development of ICR mouse one-cell embryos in the presence of minute amounts of EDTA and suggested that the beneficial role of EDTA might be related to the chelation of some metal ion(s) other than Ca^{2+} or Mg^{2+}.

Another interesting point is the paternal effect on the early development. As shown in Table 2, the rate of development beyond the 2-cell stage is significantly higher in the embryos derived from ICR eggs fertilized by C3 × B spermatozoa than in the embryos derived from ICR eggs fertilized by ICR spermatozoa. The rate, however, is much lower than that of its recip-

TABLE 3. Development beyond the 2-cell Stage of *in vitro* Fertilized Mouse Embryos Cultured in TYH Media with Different Concentrations of Sodium Pyruvate

Strain		Na-pyruvate (mM)	No. of 2-cell embryos cultured	No. (%) of embryos developed to:	
Egg	Sperm			4-cell stage in 24 h	blastocyst in 72 h
ICR	ICR	1	60	1 (1.7)	3 (5.0)
		0.5	54	5 (9.3)	2 (3.7)
		0.25	61	3 (4.9)	2 (3.3)
			175	9 (5.1)	7 (4.0)
ICR	C3 × B*	1	47	9 (19.1)	7 (14.9)
		0.5	51	6 (11.8)	8 (15.7)
		0.25	51	5 (9.8)	3 (5.9)
			149	20 (13.4)	18 (12.1)
C3 × B*	ICR	1	41	35 (85.4)	23 (56.1)
		0.5	40	37 (92.5)	34 (85.0)
		0.25	39	35 (89.7)	37 (94.9)
			120	107 (89.2)	94 (78.3)
C3 × B*	C3 × B*	1	45	29 (64.4)	24 (53.3)
		0.5	44	36 (81.8)	29 (65.9)
		0.25	42	33 (78.6)	25 (59.5)
			131	98 (74.8)	78 (59.5)

* C3 × B: C3H/He × C57BL/6J-at

rocal cross (C3 × B eggs × ICR spermatozoa). This result suggests that a paternal component might be active in the very early stages of development, probably by the 4-cell stage, although the effect is smaller than the maternal effect.

Table 3 shows the results of an experiment in which the 2-cell embryos fertilized *in vitro* were cultured in TYH medium containing various concentrations of sodium pyruvate. Again, the embryos derived from C3 × B eggs were obviously more developed than embryos derived from ICR eggs at all pyruvate concentrations. ICR eggs fertilized by F_1 sperm developed slightly better than ICR × ICR embryos especially in media containing higher levels of pyruvate.

In conclusion, the characteristics of mouse embryos obtained by fertilization *in vitro* are very similar to those of one-cell embryos fertilized *in vivo*. However, the high degree of synchrony attained by fertilization *in vitro* does not necessarily permit synchronous development thereafter. Further studies on the development of *in vitro* fertilized embryos may elucidate the genetic and physiological mechanisms controlling early development in mammals.

REFERENCES

1. Whitten, W. K. and Biggers, J. D.: Complete development *in vitro* of the preimplantation stages of the mouse in a simple chemically defined medium. *J. Reprod. Fert.*, **17**: 399–401, 1968.
2. Whitten, W. K.: Nutritional requirements for the culture of preimplantation embryos *in vitro*. *Adv. in Biosci.*, **6**: 129–139, 1971.
3. Whittingham, D. G.: *In vitro* fertilization, embryo transfer and storage. *Brit. med. Bull.*, **35**: 105–111, 1979.
4. Toyoda, Y.: Survival of frozen-thawed mouse embryos fertilized *in vitro*. In: Frozen Storage of Laboratory Animals (ed. G. H. Zeilmaker). Gustav Fischer Verlag, Stuttgart, 1981.
5. Toyoda, Y., Yokoyama, M. and Hosi, T.: Studies on the fertilization of mouse eggs *in vitro*, II: Effect of *in vitro* preincubation of spermatozoa on time of sperm penetration of mouse eggs *in vitro*. *Japan. J. Anim. Reprod.*, **16**: 152-157, 1971.
6. Toyoda, Y., Yokoyama, M. and Hosi, T.: Studies on the fertilization of mouse eggs *in vitro*, I: *In vitro* fertilization of eggs by fresh epididymal sperm. *Japan. J. Anim. Reprod.*, **16**: 147–151, 1971.
7. Kasai, K., Minato, Y. and Toyoda, Y.: Fertilization and development *in vitro* of mouse eggs from inbred strains and F_1 hybrids. *Japan. J. Anim. Reprod.*, **24**: 19-22, 1978.

8. Cross, P. C. and Brindter, R. L.: The sensitivity of one-cell mouse embryos to pyruvate and lactate. *Exp. Cell Res.*, **77**: 57–62, 1973.
9. Quinn, P. and Harlow, G. H.: The effect of oxygen on the development of preimplantation mouse embryos *in vitro*. *J. Exp. Zool.*, **206**: 73–80, 1978.
10. Motomura, M. and Toyoda, Y.: Scanning electron microscopic observations on the sperm penetration through the zona pellucida of mouse oocytes fertilized *in vitro*. *Jpn. J. Zootech. Sci.*, **51**: 595–601, 1980.
11. Oliphant, G. and Brackett, B. G.: Capacitation of mouse spermatozoa in media with elevated ionic strength and reversible decapacitation with epididymal extracts. *Fertil. & Steril.*, **24**: 948–955, 1973.
12. Fraser, L. R.: Accelerated mouse sperm penetration *in vitro* in the presence of caffeine. *J. Reprod. Fert.*, **57**: 377–384, 1979.
13. Fraser, L. R. and Quinn, P. J.: A glycolytic product is obligatory for initiation of the sperm acrosome reaction and whiplash motility required for fertilization in the mouse. *J. Reprod. Fert.*, **61**: 25–35, 1981.
14. Okamoto, M. and Toyoda, Y.: Effects of glucose and pyruvate on the sperm penetration and fertilization of mouse eggs *in vitro*. *Jpn. J. Zootech. Sci.*, **51**: 171–177, 1980.
15. Biggers, J. D., Whittingham, D. G. and Donahue, R. P.: The pattern of energy metabolism in the mouse oocyte and zygote. *Proc. Nat. Acad. Sci. U.S.A.*, **58**: 560–567, 1967.
16. Kaufman, M. H.: Timing of the first cleavage division of the mouse and the duration of its component stage: A study of living and fixed eggs. *J. Cell Sci.*, **12**: 799–808, 1973.
17. Niwa, K., Araki, M. and Iritani, A.: Fertilization *in vitro* of eggs and first cleavage of embryos in different strains of mice. *Biol. Reprod.*, **22**: 1155–1159, 1980.
18. Shire, J. G. M. and Whitten, W. K.: Genetic variation in the timing of first cleavage in mice: Effect of paternal genotype. *Biol. Reprod.*, **23**: 363–368, 1980.
19. Shire, J. G. M. & Whitten, W. K.: Genetic variation in the timing of first cleavage in mice: Effect of maternal genotype. *Biol. Reprod.*, **23**: 369–376, 1980.
20. Gamov, E. I. and Prescott, D. M.: The cell cycle during early embryogenesis of the mouse. *Exp. Cell Res.*, **59**: 117–123, 1970.
21. Luthardt, F. W. and Donahue, R. P.: DNA synthesis in developing two-cell mouse embryos. *Dev. Biol.*, **44**: 210–216, 1975.
22. Sawicki, W., Abramczuk, J. and Blaton, O.: DNA synthesis in the second and third cell cycles of mouse preimplantation development. *Exp. Cell Res.*, **112**: 119–205, 1978.
23. Abramczuk, J., Solter, D. and Koprowski, H.: The beneficial effect of EDTA on development of mouse one-cell embryos in chemically defined medium. *Dev. Biol.*, **61**: 378–383, 1977.

Discussion

Dr. Mullen: Dr. Wesley Whitten found that if he reduced the oxygen concentration in the culture system to 5% he was able to culture inbred single-cell embryos to the blastocyst stage. Have you tried reducing the O_2 concentration in your system?

Dr. Toyoda: No, I have not tried it. The culture system I used was a drop of medium under paraffin oil under 5% CO_2 in air. I think I should follow up Dr. Whitten's findings using embryos fertilized *in vitro*. But one thing I would like to emphasize is that the genetic effect might be exaggerated under such unfavorable conditions.

Dr. Zalc: Is there a clear correlation between the result of your *in vitro* fertilization and what can be observed *in vivo*? As, indeed, for the maintenance of some neurological mutant mice we have suggested the use of the same inbred strains, C3H/He and C57BL (La Chapelle *et al.*, in *Neurological Mutation Affecting Myelination* [INSERM Symposium], N. Bauman, ed., 1980).

Dr. Toyoda: Yes, ICR embryos fertilized *in vivo* showed distinctly poorer development as compared with the embryos derived from C3H/He × C57BL/6J-at F_1 females, when cultured from the pronuclear stage even under a gas mixture of 5% CO_2, 5% O_2 and 90% N_2. The interval between the first and second cleavage division *in vivo* is also longer in ICR than in F_1 embryos, and this may have a correlation with the "2-cell block" in ICR embryos *in vitro*.

Dr. Tachi: 1. Is there any possibility that the effects of maternal genetic constitution on the early development of ICR and C3FB1 ova fertilized *in vitro*, as just reported in your presentation, might be correlated with H-2 types?

2. Would you be inclined to suggest the presence of genetic control mechanisms specifically directed toward the first cleavage?

Dr. Toyoda: 1. I have not examined the H-2 type, so I have no data concerning your question.

2. No, toward the second rather than the first cleavage. First cleavage occurs in all types of eggs fertilized *in vitro*. There is a marked difference in the ability to achieve second cleavage under an unfavorable *in vitro* environment. Preparatory changes responsible for this difference may have occurred during the pronuclear stage.

Dr. Mullen: I thought ICR was a random-bred Swiss-Webster strain. Is your ICR strain inbred?

Dr. Toyoda: No, the ICR I used (JCL:ICR) is a closed colony which is maintained to keep a maximum degree of heterozygosity. So they are heterozygous at many loci. Nevertheless, development beyond the 2-cell stage is poor.

In vitro Culture of Human Follicular Oocytes

Shuetsu Suzuki

Mammalian oocytes cultured *in vitro* after liberation from follicles have been reported to mature spontaneously at a rate similar to that observed in the ovary after stimulation by gonadotropins. However, there have been relatively few reports concerning ultrastructural and biological properties of cultured human oocytes.

Recently it has been shown that preovulatory human oocytes can be fertilized by capacitated spermatozoa, and the viability of these oocytes has been confirmed by a successful birth after reimplantation.[1] The present study was designed to evaluate the maturation process of *in vitro* cultures of human follicular oocytes by electron microscopy and other methods.

MATERIALS AND METHODS

Whole ovaries or wedges of ovarian tissue were excised from patients who were laparotomized for elective gynecologic surgery. Operations were performed irrespective of the phase of the menstrual cycle and without treatment to induce ovulation. Oocytes were aspirated from each ovarian follicle and placed in droplets of modified Ham's F-10 medium for oocyte culture which contained 15% human umbilical cord serum.[2] The oocytes were cultured in Falcon plastic dishes under paraffin oil at 37° C in an atmosphere of 5% CO_2 in air.

For examination by transmission electron microscopy, oocytes were fixed at 4° C for 1 hour in 2% osmium tetroxide. After fixation, the oocytes were embedded in 3% solid agar that had been previously fixed in 2% glutaraldehyde and trimmed to approximately 8 cu mm for easier handling. Specimens were then rapidly dehydrated in a graded series of ethanol, with a final rinse in 100% acetone, then a 4:6 mixture of Epon–812 was added.[3] Silver-colored thin sections were stained with uranyl acetate

Department of Obstetrics and Gynecology, School of Medicine, Keio University, Tokyo, Japan

and lead citrate and observed in a Hitachi HU-11-B electron microscopy with accelerating voltages of 50 and 75 kv.

For examination by scanning electron microscopy, surrounding cumulus masses and zona pellucida were removed enzymatically with 0.1% pronase. After fixation by the method used for preparation for transmission electron microscopy, the specimens were dehydrated in a graded series of ethanol and placed in 100% amyl acetate. They were then mounted on square cover slips and critical point-dried. After coating with carbon and gold to a thickness of approximately 400 Å, observations and photographs were made with a JEOL scanning electron microscopy, type S-1, with an accelerating voltage of 10 kv.

In order to evaluate the biological properties of maturing human oocytes, the nuclear DNA and cytoplasmic protein content of cultured follicular oocytes was determined by cytofluorometry using a Zeiss universal microscopy with a Zeiss MPM 01 photometer head. The oocytes were stained with propidium iodide and fluorescein isothiocyanate and placed in the 25 μm space between cover slip and slide. Insemination *in vitro* with spermatozoa was performed to confirm attachment of and penetration through the zona pellucida. Ejaculated spermatozoa were washed three times with BWW medium by centrifugation (500 $\times$ g for 5 minutes) and incubated for 4 hours. They were suspended at a concentration of 2 to 3 million motile cells/ml. Oocytes were placed in a 0.1 ml droplet of the sperm suspension and cultured *in vitro* in a CO_2 incubator under the same conditions as used for the oocyte culture. Immature and cultured oocytes, surrounded by an intact zona or made zona-free by treatment with pronase, were used.

RESULTS

Ultrastructural observations

In this study, 239 oocytes were recovered from 70 patients. Fifty of these oocytes were recovered from atretic follicles and showed degenerative changes such as enlargement of the perivitelline space and cytoplasmic heterogeneity. One hundred and twenty-seven oocytes were cultured for various time periods and were used for observation and insemination. After 48 hours of incubation, 70 oocytes (55.1% of 127 oocytes) had reached metaphase of the second meiotic division (Table 1).

Immediately after recovery from the follicles, the oocytes were surrounded by thick layers of granulosa cells. Their cytoplasmic processes were projecting through the zona pellucida, which enveloped the oocytes

TABLE 1.   *In vitro* Maturation of Human Oocytes

Stage	Duration of culture		
	0 hr	24 hr	46–48 hr
Degenerating	55	0	23
Resting	48	14	3
Diakinesis	0	6	31
Metaphase II	0	0	70
Total	103	20	127

completely. The eccentrically located large germinal vesicle was spherical and had a slightly undulating membrane. In general, the organelles were distributed evenly in the ooplasm. Most of the mitochondria were spherical and had sparse cristae. The Golgi apparatus consisted of aggregates of tubuli and vesicles. Using scanning electron microscopy, numerous fine microvilli were observed on the surface of the oocytes after enzymatic removal of the zona pellucida.

Even after 27 hours of incubation, numerous microvilli were seen extending from the surface of plasma membrane into the zona pellucida. Most microvilli were single, but clearly arborizing ones were observed.

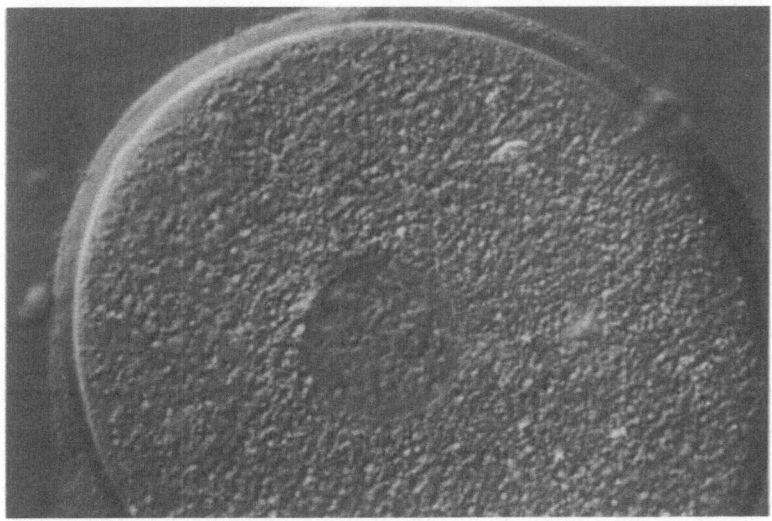

FIG. 1.   The eccentrically located large germinal vesicle is spherical in shape and has slightly undulating membrane.

Cortical granules were bounded by a single membrane and were electron-dense with a homogeneous structure. After 48 hours of incubation, the nuclear membrane had disappeared. The second-meiotic spindles and chromosomes were located near the surface of the oocytes. The regions occupied by spindles and chromosomes were relatively free of any other cytoplasmic structures. Chromosomes were located in the equatorial plate of the spindle, and the axis of the spindle was vertical to the oocyte surface.

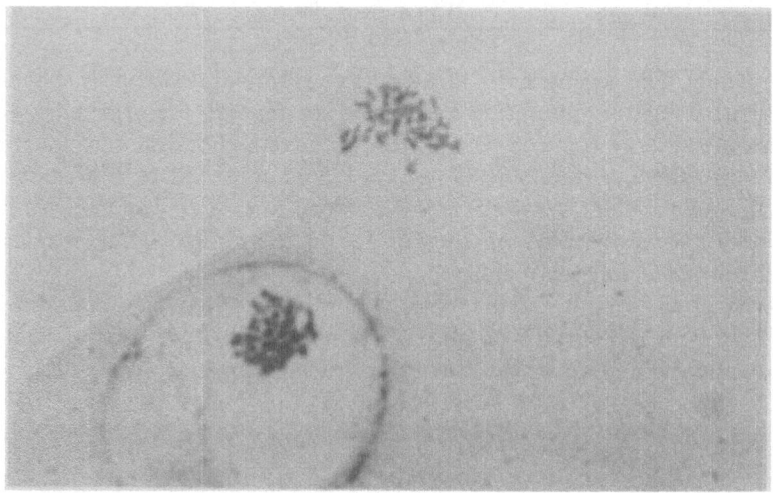

FIG. 2. After 48 hours' culture *in vitro*, the first polar body is extruded in the perivitalline space and the second metaphase chromosomes can be seen in the cytoplasm.

The first polar body had a large elliptical and irregular shape and was expelled in the perivitelline space. The surface was indented and had few microvilli. The cytoplasmic organization was similar to that of the oocyte. It contained abundant cortical granules and residual microtubuli from the first-meiotic spindle. In the perivitelline space, cytoplasmic remnants, probably derived from ooplasm, were observed. In the ooplasm, dumbbell-shaped and normal-appearing budding mitochondria were seen. The well-developed endoplasmic reticulum appeared in vacuoles of various sizes, enveloped by a limiting membrane. Most of the endoplasmic reticulum was agranular, but some granular endoplasmic reticulum was also observed. The number of cortical granules was elevated, but the Golgi apparatus was not prominent. Tubular aggregates which consisted of short and

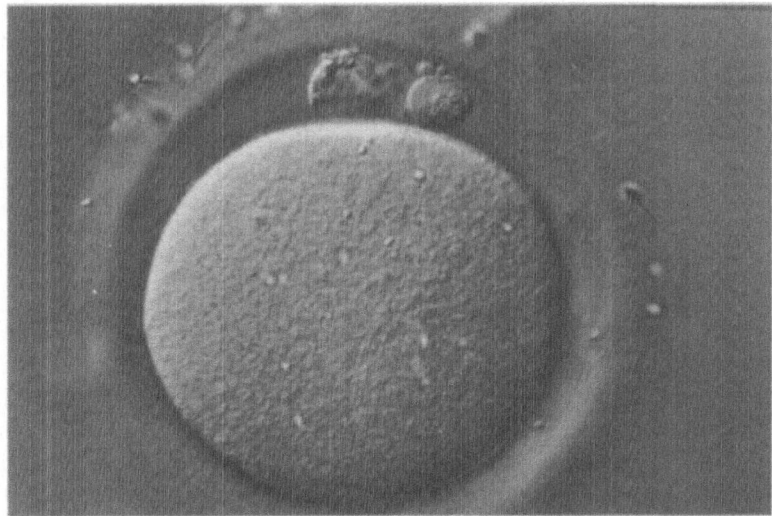

FIG. 3. After insemination by capacitated spermatozoa, two polar bodies in the perivitalline space and extra spermatozoa in the zona pellucida can be observed.

thin tubuli appeared more abundant after incubation. Microvilli of oocytes arborizing from the protruding areas of oolemma were so sparse and short that the surface of the oocyte could be seen in the interval between the microvilli.

Biological properties of cultured oocytes

The DNA and protein contents of one uncultured and four cultured oocytes were measured by cytofluorometry, and their meiotic stages were determined by fluoroscopy using propidium iodide. After 48 hours in culture and extrusion of the first polar bodies, the DNA content as measured by fluorometry was reduced to approximately half that of the germinal vesicle stage, but the protein content of the cytoplasm was not changed in spite of nuclear maturation. To obtain reproducibility of measurement, precautions were taken to ensure identical treatment of the oocytes.

After insemination, immature oocytes with germinal vesicles showed sparse sperm attachment and no evidence of penetration of the vitellus. Even after removal of the zona, these immature oocytes were not fertilized.

After 48 hours of incubation, 5 of 43 oocytes with intact zona were fertilized, and under phase-contrast microscopy 1 of the 5 appeared to have developed to the 4-cell stage. After removal of the zona, 6 of 17 oocytes which reached metaphase II after incubation revealed polyspermic penetration.

Scanning electron microscopic observations on insemination of zona-free cultured oocytes revealed spermatozoa trapped by microvilli at the head and a few portions of the tail.

DISCUSSION

In a study of monkey follicular oocytes, although the time for maturation through polar body extrusion was variable, after 46 to 48 hours in culture, post-dictyate stages of meiosis were observed in 79.7% of the oocytes. In addition, 26 of 47 ova revealed a polar body, suggesting that the nuclear stages necessary for fertilization were completed *in vitro*.[4] The fertility of such oocytes was assessed by transferring them into the fallopian tubes of inseminated recipients.[5]

Of 127 oocytes cultured in the present study, 70 reached metaphase II (55.1%). This rate seemed to correspond to rates in previous studies[6-8] in which oocytes were obtained from ovaries irrespective of the phase of the menstrual cycle and follicular size. At midcycle, human follicular oocytes are in various stages of maturation.[9] Although cultures of oocytes obtained from human menopausal gonadotropin- or clomiphene-primed human ovaries have not been reported, in pregnant mare serum gonadotropin-primed mice, 91.7% of the immature oocytes with cumulus cells showed nuclear maturation, in contrast to 59.2% of oocytes without cumulus cells.[10] These observations suggest that the meiotic potential of follicular oocytes depends both on the oocyte itself and the follicular environment. In a cytological study of oocytes obtained from adult human ovaries, 77% of the oocytes from polycystic ovaries and 86% from ovaries of patients treated with steroids were degenerating and frequently contained massive clumps of chromatin associated with the nucleolus.[11]

Reimplantation after *in vitro* fertilization made use of preovulatory oocytes obtained from large follicles (15 to 20 mm in diameter).[12] These results stress the importance of proper selection of the follicle.

In vitro fertilization as a method for the biological assessment of cultured oocytes has two main aspects: capacitation of spermatozoa and maturity of ova. The biological properties of spermatozoa, however, are evaluated by various methods using preovulatory oocytes,[13] zona-free hamster ova,[14] and salt-stored human oocytes.

As shown by this study, cultured follicular oocytes can be fertilized *in vitro* by washed spermatozoa,[15] but compared with the fertilization rate for preovulatory oocytes,[16] the rate for cultured oocytes is relatively low.[10] Thus far, ultrastructural observations have not explained the difference between oocytes matured *in vitro*[8,17] and those matured *in vivo*.[18,19] The scanning electron microscopic observations made in this study also failed to clarify this difference. Mature oocytes are thought to prevent multiple sperm penetration, decondense the sperm head, produce the female and the male pronuclei, and prepare for further development. The cortical granules under the cell surface, which increase in number with incubation, are believed to block polyspermy. However, factors such as sperm decondensation and pronuclear formation require more study and evaluation by methods other than fertilization.

This study has shown that penetration through the zona and fusion of oocytes prevent abnormal fertilization. Non-living oocytes from the ovaries of cadavers have shown sperm penetration without culture for maturation.[20] This might be due to post-mortem degeneration of the ova.

Although, in this study, male and female pronuclei were formed after the fertilization of cultured follicular oocytes, further cleavage rarely occurred. Insemination of follicular oocytes with human spermatozoa capacitated in a monkey uterus also showed a low rate of cleavage[21] and probably reflects the difference between *in vivo* and *in vitro* maturation. To clarify this difference more studies of cytoplasmic maturation are needed.

REFERENCES

1. Steptoe, P. C. and Edwards, R. G.: Birth after the reimplantation of a human embryo (letter to the editor). *Lancet*, **2**: 366, 1978.
2. Yanagimachi, R., Lopata, A., Odom, C. B., Bronson, B. A., Mahi, C. A. and Nicolson, G. L.: Retention of the biologic characteristics of zona pellucida in highly concentrated salt solution: The use of salt-stored eggs for assessing the fertilizing capacity of spermatozoa. *Fertil. Steril.*, **31**: 562, 1979.
3. Luft, J. H.: Improvements in epoxy resin embedding method. *J. Biophys. Biochem. Cytol.*, **9**: 409, 1961.
4. Suzuki, S. and Mastroianni, L., Jr: Maturation of monkey ovarian follicular oocytes *in vitro*. *Am. J. Obstet. Gynecol.*, **96**: 723, 1966.
5. Suzuki, S. and Mastroianni, L., Jr.: The fertilizability of *in vitro* cultured monkey ovarian follicular oocytes. *Fertil. Steril.*, **19**: 500, 1968.
6. Suzuki, S. and Iizuka, R.: Maturation of human ovarian follicular oocytes *in vitro*. *Experientia*, **26**: 640, 1970.
7. Edwards, R. G., Bavister, B. D. and Steptoe, P. C.: Early stages of fertiliza-

tion *in vitro* of human oocytes matured *in vitro*. *Nature* (Lond), **221**: 631, 1969.

8. Zamboni, L., Thompson, R. S. and Smith, D. M.: Fine morphology of human oocyte maturation *in vitro*. *Biol. Reprod.*, **7**: 425, 1972.

9. Moore, D. E., Thompson, R. S. and Israel, R.: Recovery of midcycle human follicular oocytes: Correlation of their morphology with endometrial and follicular histology. *Fertil. Steril.*, **29**: 518, 1978.

10. Cross, P. C. and Brinster, R. L.: *In vitro* development of mouse oocytes. *Biol. Reprod.*, **3**: 298, 1970.

11. Sanyal, M. K., Taymor, M. L. and Berger, M. J.: Cytologic features of oocytes in the adult human ovary. *Fertil. Steril.*, **27**: 502, 1976.

12. Lopata, A., Johnston, W. I. H., Hoult, I. J. and Speirs, A. I.: Pregnancy following intrauterine implantation of an embryo obtained by *in vitro* fertilization of a preovulatory egg. *Fertil. Steril.*, **33**: 117, 1980.

13. McMaster, R., Yanagimachi, R. and Lopata, A.: Penetration of human eggs by human spermatozoa *in vitro*. *Biol. Reprod.*, **19**: 212, 1978.

14. Yanagimachi, R., Yanagimachi, H. and Rogers, B. J.: The use of zona-free animal ova as a test system for the assessment of fertilizing capacity of human spermatozoa. *Biol. Reprod.*, **15**; 471, 1976.

15. Soupart, P. and Morgenstern, L. L.: Human sperm capacitation and *in vitro* fertilization. *Fertil. Steril.*, **24**: 462, 1973.

16. Lopata, A., Brown, J. B., Leeton, J. F., Talbot, J. M. and Wood, C.: *In vitro* fertilization of preovulatory oocytes and embryo transfer in infertile patients treated with clomiphene and human chorionic gonadotropin. *Fertil. Steril.*, **30**: 27, 1978.

17. Soupart, P. and Strong, P. A.: Ultrastructural observations on human oocytes fertilized *in vitro*. *Fertil. Steril.*, **25**: 11, 1974.

18. Rousseau, P., Meda, P., Lecart, C., Haumont, S. and Ferin, J.: Cortical granule release in human follicular oocytes. *Biol. Reprod.*, **16**: 104, 1977.

19. Lopata, A., Sathananthan, A. H., McBain, J. C., Johnston, W. I. H. and Speirs, A. L.: The ultrastructure of the preovulatory human egg fertilized *in vitro*. *Fertil. Steril.*, **33**: 12, 1980.

20. Overstreet, J. W. and Hombree, W. C.: Penetration of the zona pellucida of nonliving human oocytes by human spermatozoa *in vitro*. *Fertil. Steril.*, **27**: 815, 1976.

21. Seits, H. M., Rocha, G., Brackett, B. G. and Mastroianni, L.: Cleavage of human ova *in vitro*. *Fertil. Steril.*, **22**: 255, 1971.

Discussion

Dr. Beddington: If the *zona pellucida* of an ovum matured *in vivo* is removed, is there an increase in polyspermy?

Dr. Suzuki: Without the *zona pellucida*, polyspermy is very much increased, but we do not think this is caused by treatment with pronase.

Dr. Mullen: Could the pronase treatment to remove the *zona* be altering the cell membrane so that the incidence of polyspermy increases?

Dr. Suzuki: We are also slightly worried about this point, and we think the mechanical removal of the *zona* is preferred, but at this moment most researchers are using this method.

Dr. Iritani: You showed a beautiful human 4-celled egg. Did you use a preovulatory oocyte or matured oocyte from the GV stage? I ask this question because I am concerned that a matured oocyte in culture could not cleave *in vitro* even after fertilization is achieved.

Dr. Suzuki: Yes, we used the cultured ova for insemination *in vitro*. We got a few successful fertilized ova. I do not know why the cultured ova cannot be fertilized so well, but we are going to check the cytoplasmic maturation in detail.

II. GENETICS OF MAMMALIAN EMBRYOS

PROPERTIES OF MATERIALS IN PARALLEL

Molecular Insight to the Mechanisms of Cell Adhesion in Relation with Early Mammalian Morphogenesis

Tokindo S. Okada, Chikako Yoshida, Tadao Atsumi, Soh-Ichi Ogou, and Masatoshi Takeichi

INTRODUCTION

The mechanisms by which animal cells adhere to each other have long been a subject of debate.[1-3] This problem seems even more difficult to solve when the recognition of different cell types in mixtures of heterotypic cells is taken into consideration. In fact, one of the most critical problems in defining these mechanisms is explaining not only cell adhesion *per se*, but also how the cells locate their own type out of other types.

It has now been widely accepted that the mechanisms underlying cell adhesion and cell recognition provide the basis for morphogenesis of the multicellular hierarchy of developing animals.[3,4] The involvement of ligand or adhesive molecules in cell adhesion and recognition has been presumed, and the identification and isolation of such hypothetical molecules has been reported by several laboratories.[5-8] Most of these efforts have looked for the active molecule(s) in the intercellular exudate, but not on the cell surfaces directly. Through a series of works conducted in our laboratory over the past decade, we now believe that we have identified and isolated adhesive molecule(s) located on the cell surfaces.[9-11] We therefore propose a new hypothesis of cell adhesion, "dual adhesion sites," which more accurately explains these mechanisms.

The main objectives of this communication are to review this hypothesis and to present our recent efforts to apply the hypothesis to the study of early mammalian embryogenesis using embryonal carcinoma cells as well as early embryonic cells.

DUAL ADHESION SITES HYPOTHESIS

An essential point of this hypothesis is the assumption that tissue cells

Institute for Biophysics, Faculty of Science, Kyoto University, Kyoto, Japan utilize two distinctly different mechanisms for their contact. The mechanisms involve two different classes of adhesion molecules located on cell surfaces. One class requires Ca^{2+} to function whereas the other does not. They have been designated CDS (Ca^{2+}-dependent site) and CIDS (Ca^{2+}-independent site).[12,13]

By treating cells under strictly controlled conditions, it is now possible to leave either of these two sites intact on the cell surfaces. For example, when cells were dissociated with 0.01% trypsin solution *containing Ca^{2+}* (TC-treatment in Fig. 1), only the CDS was retained and the aggregation occurred only in the medium with Ca^{2+}. Cells dissociated with 0.0001%

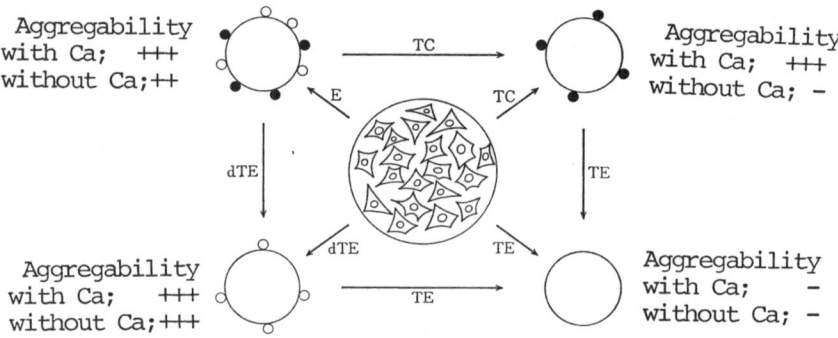

FIG. 1. Experimental removal of CDS (shown with ●) and/ or CIDS (shown with ○) by dissociating cells with specificed treatment. E; treatment with EDTA, TE; treatment with 0.01% trypsin and EDTA. See text for TC and dTE. (From Takeichi, 1978)[10]

trypsin *without Ca^{2+}* (dTE-treatment in Fig. 1) retained only CIDS and their aggregation occurred in Ca^{2+}-free medium as well (Fig. 1). Thus it is assumed that CDS is very sensitive to trypsin but is resistant to this enzyme in the presence of Ca^{2+}. In contrast, CIDS, though less sensitive to trypsin than CDS, is not protected by Ca^{2+}.[14,15]

Several observations have been reported which suggest the involvement of these adhesive sites in cellular recognition. When treated cells which retain only CIDS are mixed with those retaining only CDS, distinct sorting occurs even among cells of the same type. On the other hand, no selectivity in adhesion was observed among different cell types, if the cells were similarly treated to leave only either CDS or CIDS intact.[14,16]

The dual adhesion sites hypothesis was originally based on extensive studies using Chinese hamster fibroblast cells (line V79). The validity of this hypothesis has been tested using several other cell lines as well as some freshly dissociated tissue cells of avians and mammalians.[14,16-18] Recently, our results have been confirmed by several other laboratories.[19-23]

Finally, by combining immunochemical studies using specific antibodies against each adhesive site with electrophoresis of radio-iodinated cell surface components, CIDS was identified as a protein of 125K daltons.[24] Although a specific antibody test has not yet been performed, a search for CDS by electrophoresis of labeled proteins has shown that a protein with a molecular weight of 150K daltons is a candidate for this adhesive site. With this in mind, we have initiated a study to test this hypothesis using mouse embryonal carcinoma cells.

MOLECULAR BASIS OF SELECTIVE ADHESION BETWEEN EMBRYONAL CARCINOMA CELLS AND DIFFERENTIATED CELLS

Involvement of CIDS in the selective adhesion

The multipotent embryonal carcinoma (EC) cells (line AT805)[25] used in this study was established from embryoid body line SSEB[26] originally derived from Stevens' OTT6050.[27] When the tumors developed in histocompatible mice by injecting AT805 cells were dissociated and cultured in vitro, the stem cells of EC formed islets encircled by differentiated cells. The border between these two cell populations was very distinctly demarcated, suggesting the occurrence of selective adhesion between these two cell types. When mixed aggregates of EC cells and of lined fibroblastic cells (FBL) were transferred into stationary cultures, they soon attached to the culture substrate and began to multiply. A clear separation of these two cell types was observed in such cultures, as shown in Fig. 2.

It was shown that EC cells, AT805 cells, and nullipotent EC cell line have both CDS and CIDS, as do all FBL lines including Chinese hamster V79, mouse W3, L, and 129F cells. The 129F line was established in our laboratory from a mass culture of fetal dermal cells of 129/Sv strain mice which are syngenic to EC cells. The question then arises, which site, CDS or CIDS, is involved in the recognition of EC and FBL cells. Due to technical difficulties, the studies of CIDS have not yet been completed. However, there is little doubt that CDS has a function in specific cell adhesion.

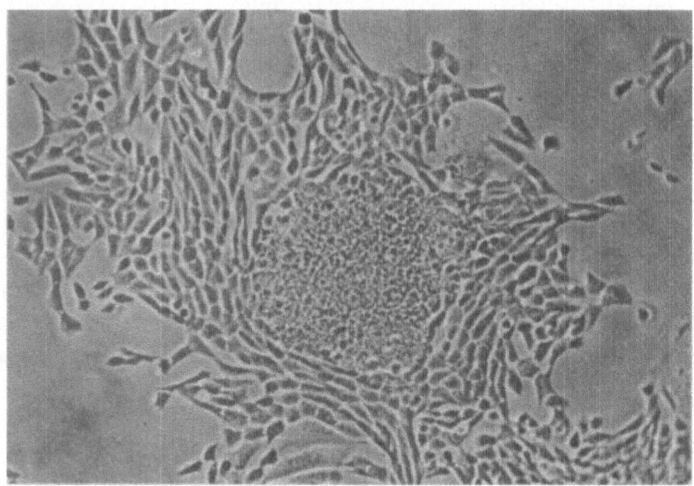

FIG. 2. A heterotypic aggregate formed by mixing AT805
cells (EC) and V79 cells (FBL) for 24 hours under gyration was
plated on the substrate in a stationary culture and incubated
for 18 hours. AT805 cells are located internally and sharply
separated from surrounding fibroblasts. ($\times$ 70)

This was demonstrated by assaying the selective adhesion in mixtures
of all possible binary combinations of the six lines mentioned above.[28]
Cell suspensions were prepared by treatment with trypsin (or pronase)
containing Ca²⁺ which leaves only CDS intact. Cells of either of binary
combination were stained with fluorescein-isothiocyanate (FITC) and
mixed aggregates formed after one hour of incubation under gyration were
observed under a fluorescent microscope (Fig. 3). The results clearly
demonstrate the occurrence of selective adhesion between EC and FBL
cells (Figs. 3A and B), while cells belonging to the same type aggregate
randomly (Figs. 3C and D).

Molecular heterogeneity of CDS

The molecular basis of cell specificity in adhesion by CDS was explored
using antibodies raised against F9 cells, which were dissociated leaving
only CDS intact. Fab fragments of these antibodies completely inhibited
the aggregation of AT805 and F9 cells, but not of FBL cells. This inhibitory effect of Fab was removed by the addition of EC cells with CDS,
but not with FBL cells. However, if EC cells having only CIDS intact were

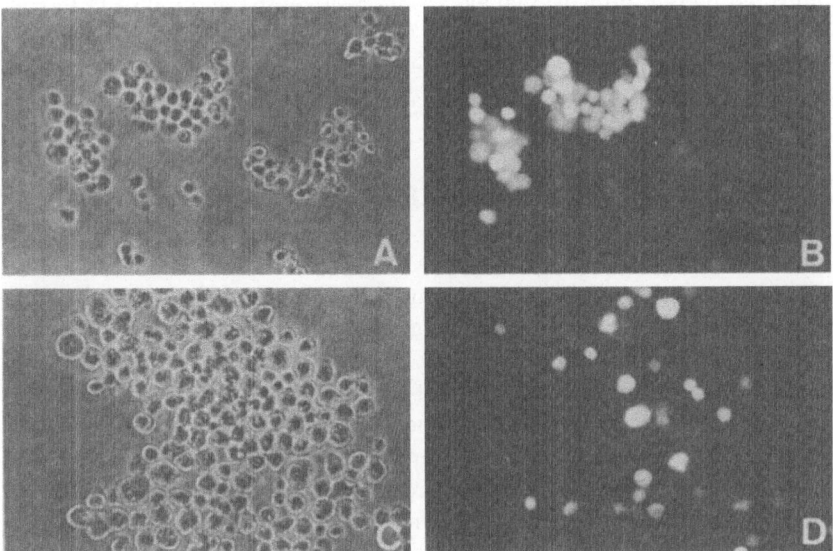

FIG. 3.   Selective and non-selective adhesion in mixed aggregates. A and B; distinct selective adhesion occurs between F9 cells (labeled with FITC) and V79 cells: C and D; Random adhesion occurs between V79 cells (labeled with FITC) and 129F cells. B and D are fluorescent micrographs of the same fields as shown in A and C (phase-contrast micrographs), respectively. (× 80)

added, Fab still retained its inhibitory effect. From these results, we can conclude that adhesion between EC cells is mediated by CDS molecules that are antigenetically different from the molecules which bind FBL cells. The molecular basis of selective adhesion between EC and FBL cells, therefore, must be examined for the differences in molecular species of CDS in different cell types.

The identification of two different species of CDS molecules was accomplished using SDS-polyacrylamide gel electrophoresis of radio-iodinated surface components. In FBL cells, it was shown that a component with a molecular weight of 150,000 (p150K) was correlated specifically with CDS. This component was detected in samples prepared from cells having only CDS intact, but not in samples treated to remove CDS from their surfaces. In EC cells, a candidate of CDS is not a 150K component, but a protein with a molecular weight of 140,000 (p140K). These results were confirmed by two-dimensional electrophoresis by the method of O'Farrell and O'Farrell.[29]

It is of special interest from the viewpoint of embryology to see if CDS and/or CIDS are functioning in the early embryonic period. We also hope to determine which molecular species of CDS (p150K or p140K) appears.

MOLECULAR NATURE OF ADHESIVE SITES OF EARLY MOUSE EMBRYOS

The assay system for identifying the adhesive sites of early embryonic cells consisted of examining the aggregates (embryo fusion) of several embryos and of embryos and EC or FBL cells placed in a Petri dish on a gyratory shaker. Since the number of early embryonic cells available was very limited, the aggregation of dissociated cells was not assayed. Mouse embryos ranging from one cell to morula stages were denuded[30] and subjected to three different types of treatment: (i) high concentration of trypsin with Ca^{2+} (leaving only CDS intact); (ii) low concentration of trypsin without Ca^{2+} (leaving only CIDS intact); and (iii) high concentration of trypsin with EDTA (removing both CDS and CIDS). It was soon disclosed that, in the aggregation of early embryos, CIDS does not seem to display a significant function. However, embryos subjected to treatment (i) (leaving only CDS intact) aggregated efficiently in the medium with Ca^{2+}, but not in the medium without Ca^{2+}. Thus, early embryonic cells have adhesive sites which require Ca^{2+} (i.e., CDS).

The aggregation of early embryos presumably mediated by CDS was completely inhibited by Fab fragments of antibodies of F9 cells. The inhibitory effect of Fab was removed by EC cells (both AT805 and F9 cells) subjected to treatment (i) to leave CDS intact. The effect was not removed by: (a) EC cells subjected to treatment (ii) or (iii) to remove CDS; (b) FBL cells dissociated by any one of the treatments described earlier. These results demonstrate that the CDS of early embryos is the same as the molecule found in EC cells (i.e., p140K protein). In fact, EC cells with CDS intact adhered to embryos treated as in (i), whereas FBL cells with CDS intact did not.

There are several indications that p150K exists on the cell surfaces of embryonic nervous tissues and several lined tumor cells. Perhaps, p150K can be called *late* CDS, while p140K is *early* CDS. Thus, there must be a molecular heterogeneity of CDS and the shift from the *early* to the *late* molecules must occur in the ontogenic process. As shown, selective adhesion occurs between cells with different CDS molecules. Then, it is assumed that the shift from the *early* molecules to the *late* molecules is associated with profound changes in morphogenesis related to cellular recognition. It is of interest to determine when and in which parts of the embryo this molecular shift occurs. On the other hand, we must determine whether p140K and p150K are coded by two different genes.

Acknowledgements

The work reviewed here has been supported by a Grant for Basic Cancer Research from the Japan Ministry of Education, Science and Culture to T. S. O. and M. T. Details of techniques as well as of experimental results have been given in separate original papers (including those in preparation) from our laboratory. We thank Mrs. Yonesaki for helping us in the preparation of this manuscript.

REFERENCES

1. Curtis, A. S. G.: The Cell Surface: Its Molecular Role in Morphogenesis. Logos Press, London, 1967.
2. Weiss, L.: The Cell Periphery, Metastasis and Contact Phenomena. North-Holland, Amsterdam, 1967.
3. Trinkaus, J. P.: Cells into Organs: The Forces That Shape the Embryo. Prentice-Hall, Englewood Cliffs, New Jersey, 1969.
4. Townes, P. L. and Holtfreter, J.: Directed movements and selective adhesion of embryonic amphibian cells. *J. Exptl. Zool.*, **128**: 53–120, 1955.
5. Hausman, R. E. and Moscona, A. A.: Isolation of retina-specific cell-aggregating factor from membranes of embryonic neural retina tissue. *Proc. Nat. Acad. Sci. U.S.A.*, **73**: 3594–3598, 1976.
6. Kudo, K., Hanaoka, Y. and Hayashi, H.: Characterization of tumor cell aggregation promoting factor from rat ascites hepatoma cells: Separation of two factors with different antigenic property. *Br. J. Cancer*, **33**: 79–90, 1976.
7. Thiery, J-P., Brackenbury, R., Rutishauser, U. and Edelman, G. M.: Adhesion among neural cells of chick embryos, II: Purification and characterization of a cell adhesion molecule from neural retina. *J. Biol. Chem.*, **252**: 6841–6845, 1977.
8. Bertolotti, R., Rutishauser, U. and Edelman, G. M.: A cell surface molecule involved in aggregation of embryonic liver cells. *Proc. Nat. Acad. Sci. U.S.A.*, **77**: 4831–4835, 1980.
9. Okada, T. S., Takeichi, M., Yasuda, K. and Ueda, M. J.: The role of divalent cations in cell adhesion. In: Advances in Biophysics, Vol. 6 (ed. M. Kotani), p. 157. Japan Scientific Societies Press, Tokyo, 1976.
10. Takeichi, M.: Recognition in cell adhesion. *Kagaku*, **48**: 474–481, 1978.

11. Takeichi, M.: Identification of cell-to-cell adhesion molecules of Chinese hamster fibroblasts. In: Cancer Cell Biology (ed. T. Nagayo and W. Mori), Gann Monograph on Cancer Research 25, pp. 3–8. Japan Scientific Societies Press, Tokyo, 1981.

12. Takeichi, M.: Functional correlation between cell adhesive properties and some cell surface proteins. *J. Cell. Biol.*, **75**: 464–474, 1977.

13. Urushihara, H., Ueda, M. J., Okada, T. S. and Takeichi, M.: Calcium-dependent and -independent adhesion of normal and transformed BHK cells. *Cell Struct. Funct.*, **2**: 289–296, 1977.

14. Takeichi, M., Ozaki, H. S., Tokunaga, K. and Okada, T. S.: Experimental manipulation of cell surface to affect cellular recognition mechanisms. *Develop. Biol.*, **70**: 195–205, 1979.

15. Atsumi, T. and Takeichi, M.: Cell association pattern in aggregates controlled by multiple cell-cell adhesion mechanisms. *Develop. Growth and Differ.*, **22**: 133–142, 1980.

16. Urushihara, H., Ozaki, H. S. and Takeichi, M.: Immunological detection of cell surface components related with aggregation of Chinese hamster and chick embryonic cells. *Develop. Biol.*, **70**: 206–216, 1979.

17. Aoyama, H., Okada, T. S. and Takeichi, M.: Analysis of the cell adhesion mechanism using somatic cell hybrids: I. Aggregation of hybrid cells between adhesive V79 and non-adhesive Ehrlich's ascites tumor cells. *J. Cell Sci.*, **43**: 391–406, 1980.

18. Ueda, K., Takeichi, M. and Okada, T. S.: Differences in the mechanisms of cell-cell and cell-substrate adhesion revealed in a human retinoblastoma cell line. *Cell Struct. Funct.*, **5**: 183–190, 1980.

19. Grunwald, G. B., Geller, R. L. and Lillien, J.: Enzymatic dissection of embryonic cell adhesive mechanisms. *J. Cell Biol.*, **85**: 766–776, 1980.

20. Magnani, J. L., Thomas, W. A. and Steinberg, M. S.: Two distinct adhesion mechanisms in embryonic neural retina cells. I. A kinetic analysis. *Develop. Biol.*, **81**: 96–105, 1981.

21. Thomas, W. A. and Steinberg, M. S.: Two distinct adhesion mechanisms in embryonic neural retina cells, II: An immunological analysis. *Develop. Biol.*, **81**: 106–114, 1981.

22. Thomas, W. A., Thomson, J., Magnani, J. L. and Steinberg, M. S.: Two distinct adhesion mechanisms in embryonic neural retina cells, III: Functional specificity. *Develop. Biol.*, **81**: 379–385, 1981.

23. Brackenbury, R., Rutishauser, U. and Edelman, G. M.: Distinct calcium-independent and dependent adhesion systems of chicken embryo cells. *Proc. Nat. Acad. Sci. U.S.A.*, **78**: 387–391, 1981.

24. Urushihara, H. and Takeichi, M.: Cell-cell adhesion molecules: Identification of a glycoprotein relevant to the Ca^{2+} -independent aggregation of Chinese hamster fibroblasts. *Cell*, **20**: 363–371, 1981.

25. Atsumi, T. and Uno, K.: Clonal teratocarcinoma stem cells have similar

adhesion mechanisms to cells from differentiated tissues. *Cell Struct. Funct.*, **4**: 388, 1979.

26. Amano, S., Uno, K. and Hagiwara, A.: Cardiac muscle cell differentiation *in vitro* from a characteristic cell line isolated from mouse teratocarcinoma. *Develop. Growth and Differ.*, **20**: 41–47, 1978.

27. Stevens, L. C.: The development of transplantable teratocarcinomas from intratesticular grafts of pre- and postimplantation mouse embryos. *Develop. Biol.*, **21**: 364–382, 1970.

28. Takeichi, M., Atsumi, T., Yoshida, C., Uno, K. and Okada, T. S.: Selective adhesion of embryonal carcinoma cells and differentiated cells by Ca^{2+}-dependent sites. *Develop. Biol.*, **87**: 340–350, 1981.

29. O'Farrell, P. H. and O'Farrell, P. A.: Two-dimensional polyacrylamide gel electrophoretic fractionation. In: Methods in Cell Biology (ed. G. Stein, J. Stein and L. J. Kleinsmith), 16, pp. 407–420. Academic Press, Inc., New York, 1977.

30. Nicholas, J. S. and Hall, B. V.: Experiments on developing rats, II: The development of isolated blastomeres and fused eggs. *J. Exptl. Zool.*, **90**: 441–459, 1942.

Discussion

Dr. Zalc: Does your anti-F9 antibody react with the 140K protein? Might there not be cross-reactivity between your anti-F9 and Forsman hapten?

Dr. Okada: We do not think that our anti-F9 is directed to the Forsman antigen, because absorption of anti-F9 with several tissues with Forsman antigen did not remove the reactivity with antigens in teratocarcinoma cells.

Dr. Le Douarin: Does your antibody raised against F9 cells inhibit compaction of the mouse embryo?

Dr. Okada: Yes, our anti-F9 Fab inhibits compaction of mouse embryos.

Genetic Influences on Teratocarcinogenesis in Mice*

Leroy C. Stevens

INTRODUCTION

Teratomas have become interesting to developmental biologists because their undifferentiated embryonic stem cells, called embryonal carcinoma cells, have been found to be equivalent to normal totipotent embryonic ectodermal cells of the 6-day egg cylinder. They have been shown to be similar morphologically,[1] antigenically,[2,3] biochemically,[4,5] and in developmental potential.[6-11] Embryonal carcinoma cells, may be easily obtained in large numbers from transplantable teratocarcinomas and established cell lines.

Teratomas are tumors composed of many kinds of chaotically arranged cells and tissues (Fig. 1). If all of the cells are differentiated, the tumors are benign and are referred to as teratomas. All cells and tissues in teratomas are derived from embryonal carcinoma cells. If the tumors are composed of well differentiated cells and embryonal carcinoma cells, they are referred to as teratocarcinomas, and may be malignant. If the tumors are composed only of embryonal carcinoma cells, they are referred to as embryonal carcinomas. All three varieties may be included in the term teratoma.

Neural tissue occurs in all primary teratomas in mice, and is one of the first tissues to differentiate.[12] Neural tissue is also one of the first tissues to differentiate in the normal development of the mouse. In addition to neural tissue, cartilage, muscle, bone, glandular tissue, skin notochord, and various kinds of epithelia are commonly seen.[13] Teeth, lens, islets of Langerhans, and gonadal tissue have rarely been observed. Tissues such as kidney, lung, and liver have never been observed in primary or transplantable teratocarcinomas, even though it has been shown that embryonal carcinoma cells can contribute to the formation of these organs in chimeric mice.[8,9] In fact, embryonal carcinoma cells are totipotent, and can dif-

The Jackson Laboratory, Bar Harbor, Maine, U.S.A.
* The research of the author reported here was supported by research grant CA 02662 from the U.S. National Institutes of Health. The Jackson Laboratory is fully accredited by the American Association for Accreditation of Laboratory Animal Care.

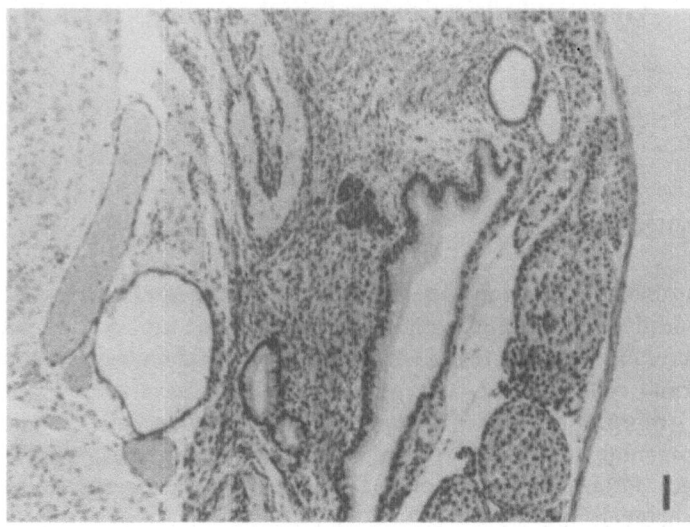

FIG. 1. Testicular teratoma with neural tissue (left), cardiac muscle and glandular tissue (middle), and endodermal epithelium. Normal testicular tubules to the right.

ferentiate into all kinds of cells and tissues including functional sperm and eggs.[8,10]

Perhaps the relative rarity of some kinds of tissues in teratomas reflects the degree of complexity of events involved in triggering their differentiation. For example, if a differentiation were the result of the interaction of two of more kinds of cells, the probability of these cells being in juxtaposition in a tumor would be remote.

Perhaps the prevalence of neural tissue in teratomas is because the early development of these tumors is similar to the development of a normal embryo[14,15] (Figs. 2–9). Both ovarian and testicular teratomas are derived from germ cells. Ovarian teratomas are derived from eggs that have undergone the first meiotic division, and begin to develop parthenogenetically within the ovary.[16] They undergo apparently normal cleavage, blastocyst, and even egg cylinder formation.[17] The morphology of the egg cylinder (five to six-day embryo) is simple. A layer of ectodermal epithelium surrounds a proamniotic cavity. The ectoderm is surrounded by a single cell layer of primary endoderm. The transformation of ectoderm to immature neuroepithelium occurs very rapidly during the 8th day of gestation. Apparently, in both teratomatous and normal development, the major requirement for the differentiation of neuroepithelium is the formation of a proamniotic cavity by totipotent ectodermal cells.

The early development of testicular teratomas also resembles normal development. They are derived from primordial germ cells that transform into embryonal carcinoma cells at about 12 days gestation.[18] They are first recognizable within the seminiferous tubules in embryos 15 days of gestation. They appear as small clusters of embryonal carcinoma cells. They very soon become epithelial and form cavities like the proamniotic cavity of normal embryos.

The early development of testicular teratomas differs from the development of ovarian teratomas and normal embryos. During normal development, the primary endoderm first appears as a single layer of cells on the ventral surface of the inner cell mass of blastocysts 4 1/2 days of gestation. The endodermal cells are determined by their location and their exposure to the fluid filled blastocoel cavity.[19,20] There is no blastocoelic cavity in early testicular teratomas, and primary endoderm fails to develop.

As in normal and ovarian teratoma development, one of the first morphogenetic events in the development of a testicular teratoma is the formation of a proamniotic cavity. Shortly after the formation of the proamniotic cavity, neural differentiation occurs. The early appearance of neural tissue in testicular teratomas suggests that an interaction between embryonic ectoderm and primary endoderm is not required for neural differentiation. This interpretation is supported by the fact that when the embryonic ectoderm is separated from the primary endoderm of 6-day embryos, it can give rise to all of the cell types observed in teratomas.[21]

MATERIALS AND METHODS

Several genetic and environmental factors are involved in determining susceptibility to testicular teratocarcinogenesis. These tumors are common only in strain 129 indicating a genetic basis for susceptibility. The incidence of teratomas varies among different sublines of strain 129 mice, further indicating genetic influence.

Several single genes have been found to influence susceptibility to teratocarcinogenesis.

Mice heterozygous for the gene steel (Sl) have twice as many teratomas as their normal Sl/+ and +/+ littermates.[22] Teratomas cannot be experimentally induced in mice homozygous for steel (Sl/Sl) because they lack primordial germ cells which give rise to these tumors.

Alleles of the agouti locus have a strong influence on suceptibility to

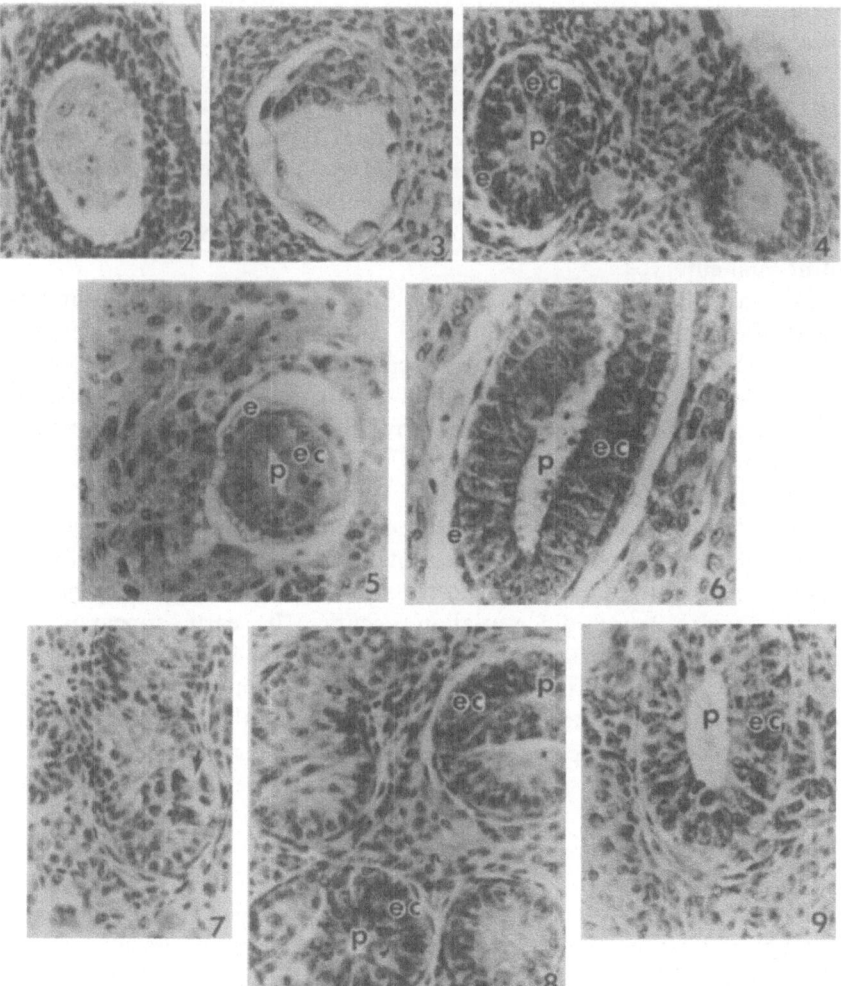

FIG. 2. Parthenogenetic morula in the ovary of a strain LT mouse.

FIG. 3. Parthenogenetic blastocyst in the ovary of a strain LT mouse.

FIG. 4. Parthenogenetic egg cylinder in the ovary of a strain LT mouse. Endoderm (e), ectoderm (ec), and proamniotic cavity (p). Normal egg to the right.

FIGS. 5 and 6. Normal 5- and 6-day egg cylinder. Endoderm ▶

teratocarcinogenesis.[22] Mice heterozygous for yellow (A^y) have fewer tumors than their $+/+$ littermates.

Teratomas occur twice as frequently in the left as in the right testis. We introduced the gene situs inversus viscerum (iv) onto the strain 129 genetic background to see if it would influence this asymmetry.[22] About half of the offspring of iv/iv × iv/iv matings are normal, and the other half have left-right transposition of the viscera including the origin of the spermatic artery. In normal mice the origin of the right spermatic artery is anterior to that of the left. Those offspring with normally situated viscera had twice as many teratomas in the left testis as in the right. Those with reversed viscera had twice as many tumors in the right as in the left testis. It is possible that the level of origin of the spermatic artery or factors associated with it influences teratocarcinogenesis.

Another gene, ter, increases the incidence of testicular teratomas from about 5% to about 33%. A single mating of strain 129/Sv-$W C P$ produced 38 male offspring, and 8 of them had teratomas. This is much higher than the general incidence of tumors in that colony. A new colony derived from that mating was established. It is designated strain 129/Sv-ter and the incidence of teratomas in this colony has remained 33%.

Genetic factors are also involved in susceptibility to experimentally induced teratocarcinogenesis. Testicular teratomas can be experimentally induced in several inbred strains of mice.[23] This was unexpected considering that these tumors are so rare that, except for two cases, the only testicular teratomas that have been reported in the literature have been those from my laboratory.

The method of experimentally inducing testicular teratomas in inbred mice is simple. When the gonadal primordium, the genital ridges, are dissected from 12-day fetuses, and grafted to adult testes, they develop into testes, and most of them have several foci of teratomas. These tumors are initiated almost immediately and can be recognized histologically in the

◀ (e), ectoderm (ec), and proamniotic cavity (p).

FIG. 7.   Cluster of embryonal carcinoma cells within a seminiferous tubule of a 16-day fetal mouse (arrow).

FIG. 8.   Two early teratomas within the seminiferous tubules of a 17-day fetal mouse. Ectoderm (e) and proamniotic cavity (p).

FIG. 9.   Early teratoma in the testis of a 18-day fetal mouse. Note similarity to normal egg cylinder.

grafts as early as six days after the genital ridges are transplanted. They develop in the same manner as spontaneous teratomas. Primordial germ cells within the seminiferous tubules begin to develop as a result of a process similar to parthenogenesis. They are first recognized as clusters of embryonal carcinoma cells that resemble cells of the inner cell mass of the blastocyst stage (Figs. 7–9). They soon develop cavities similar to the proamniotic cavity in egg cylinders. The embryonal carcinoma cells then become disorganized and form tumors composed of many kinds of haphazardly arranged cells and tissues.

RESULTS AND DISCUSSION

This simple method of inducing teratomas enabled us to investigate their cell of origin.[18] The gene steel was introduced onto the strain 129 genetic background. Fetuses homozygous for the gene steel (*Sl/Sl*) lack or have very few primordial germ cells and they die after 12 days of gestation. When genital ridges from 12-day fetuses resulting from *Sl/+* × *Sl/+* matings were grafted to adult testes, 75% of the normal (*Sl/+* and +/+) genital ridges developed into testes with teratomas. This is the same incidence as for strain 129. None of the genital ridges from their *Sl/Sl* littermates developed into testes with teratomas. *Sl/Sl* fetuses lack pigment cells. The genotypes of the offspring of the *Sl/+* × *Sl/+* matings were identified by grafting a piece of dorsal skin along with the genital ridges. *Sl/+* and +/+ grafts grew black hairs whereas their *Sl/Sl* grafts had white hairs.

The method of experimentally inducing testicular teratomas also enabled us to tentatively identify a gene lacking in strain 129 that enhances susceptibility to teratocarcinogenesis.[24] Two strains, 129 and A/He, are more susceptible to experimental induction of teratomas than any other strain tested. About 80% of grafted strain 129, and about 70% of strain A/He ridges developed into testes with teratomas. Ninety-seven percent of grafted genital ridges from 129 × A/He F₁ hybrids developed into testes with teratomas, a higher incidence than either parent strain. This very high susceptibility of hybrid genital ridges was unexpected since we have autopsied tens of thousands of hybrids and have never seen a spontaneous testicular teratoma. Apparently there is a gene in strain A/He that, in combination with genes in strain 129, increases susceptibility to teratocarcinogenesis.

There is another line of evidence that suggests that strain A/He has a gene lacking in 129 that influences susceptibility to experimental teratocarcinogenesis. We made 11 recombinant inbred strains using strains 129 and A/He as progenitors.[23] About half of them were highly susceptible, as were the parent strains. Two of them, however, were more susceptible than either parent strain. Apparently, a gene from A/He was fixed in combination with the genes of strain 129 origin that enhances susceptibility to experimental teratocarcinogenesis.

Mutant genes have been used to analyze many neoplastic and developmental phenomena in the mouse. Knowledge of the mouse genome is expanding rapidly, and analyses of more phenomena in greater depth can be expected.

REFERENCES

1. Pierce, G. B., Stevens, L. C. and Nakane, P. K.: Ultrastructural analysis of the early development of teratocarcinomas. *J. Nat. Cancer Inst.*, **39**: 755–773, 1967.
2. Edidin, M., Patthey, H. L., McGuire, E. J. and Sheffield, W. D.: An antiserum to "embryoid body" tumor cells that reacts with normal mouse embroys. In: Embryonic and Fetal Antigens in Cancer (ed. N. G. Anderson and J. H. Coggin, Jr.), p. 239. Oak Ridge National Laboratory, Oak Ridge, Tennessee, 1971.
3. Artzt, K., Dubois, P., Bennett, D., Condamine, H., Babinet, C. and Jacob, F.: Surface antigens common to mouse cleavage embryos and primitive teratocarcinoma cells in culture. *Proc. Nat. Acad. Sci. U.S.A.*, **70**: 2988–2992, 1973.
4. Damjanov, I., Solter, D. and Skreb, N.: Enzyme histochemistry of experimental embryo-derived teratocarcinomas. *Z. Krebsforsch.*, **76**: 249–256, 1971.
5. Bernstine, E. G., Hooper, M. L., Grandchamp, S. and Ephrussi, B.: Alkaline phosphatase activity in mouse teratomas. *Proc. Nat. Acad. Sci. U.S.A.*, **70**: 3899–3903, 1973.
6. Kleinsmith, L. J. and Pierce, G. B.: Multipotentiality of single embryonal carcinoma cells. *Cancer Res.*, **24**: 1544–1552, 1964.
7. Brinster, R. L.: The effect of cells transferred into the mouse blastocyst on subsequent development. *J. Exp. Med.*, **140**: 1049–1056, 1974.
8. Mintz, B. and Illmensee, K.: Normal genetically mosaic mice produced from malignant teratocarcinoma cells. *Proc. Nat. Acad. Sci. U.S.A.*, **72**: 3585–3589, 1975.
9. Illmensee, K. and Mintz, B.: Totipotency and normal differentiation of single teratocarcinoma cells cloned by injection into blastocysts. *Proc. Nat. Acad. Sci. U.S.A.*, **73**: 549–553, 1976.
10. Illmensee, K.: Reversion of malignancy and normalized differentiation of

teratocarcinoma cells in chimeric mice. In: Gatlinburg Symposium on Genetic Mosaics and Chimeras in Mammals (ed. L. B. Russell), p. 3. Plenum Press, New York, 1978.

11. Papaioannou, V. E., Gardner, R. L., McBurney, M. W., Babinet, C. and Evans, M. J.: Participation of cultured teratocarcinoma cells in mouse embryogenesis. *J. Embryol. Exp. Morphol.*, **44**: 93–104, 1978.

12. Stevens, L. C.: Embryology of testicular teratomas in strain 129 mice. *J. Nat. Cancer Inst.*, **23**: 1249–1295, 1959.

13. Stevens, L. C. and Hummel, K. P.: A description of spontaneous congenital testicular teratomas in strain 129 mice. *J. Nat. Cancer Inst.*, **18**: 719–747, 1957.

14. Stevens, L. C.: Comparative development of normal and parthenogenetic mouse embryos, early testicular and ovarian teratomas, and embryoid bodies. In: Teratomas and Differentiation (ed. M. I. Sherman and D. Solter), p. 17. Academic Press, Inc., New York, 1975.

15. Stevens, L. C.: Teratocarcinogenesis and spontaneous parthenogenesis in mice. In: Results and Problems in Cell Differentiation, Vol. 11, Differentiation and Neoplasia (Ed. R. G. McKinnel), p. 265. Springer Verlag, Berlin, 1980.

16. Eppig, J. J., Kozak, L. P., Eicher, E. M. and Stevens, L. C.: Ovarian teratomas in mice are derived from oocytes that have completed the first meiotic division. *Nature*, **269**: 517–518, 1977.

17. Stevens, L. C. and Varnum, D. S.: The development of teratomas from parthenogenetically activated ovarian mouse eggs. *Develop. Biol.*, **37**: 369–380, 1974.

18. Stevens, L. C.: Origin of testicular teratomas from primordial germ cells in mice. *J. Nat. Cancer Inst.*, **38**: 549–552, 1967.

19. Gardner, R. L. and Johnson, M. H.: Investigation of cellular interaction and development in the mammalian embryo using interspecific chimaeras between the rat and mouse. In: Cell Patterning: Ciba Foundation Symposium 29, p. 183. Elsevier/Excerpta Medica, Amsterdam, 1975.

20. Rossant, J. R.: Investigation of the determinative state of the mouse inner cell mass. II. The fate of ioslated inner cell masses transferred to oviduct. *J. Embryo. Exp. Morphol.*, **33**: 991–1001, 1975.

21. Diwan, S. B. and Stevens, L. C.: Development of teratomas from the ectoderm of mouse egg cylinders. *J. Nat. Cancer Inst.*, **57**: 937–942, 1976.

22. Stevens, L. C.: Genetic influences on teratocarcinogenesis in mice. In: A Century of Mammalian Genetics and Cancer, 1929–2029, View in Midpassage (ed. E. S. Russell). Alan R. Liss, Inc., New York, 1980.

23. Stevens, L. C.: Unpublished.

24. Stevens, L. C.: Experimental production of testicular teratomas in mice of strains 129, A/He, and their F_1 hybrids. *J. Nat. Cancer Inst.*, **44**: 923–929, 1970.

Phenotypical Expression in Nude ↔ Normal Chimeric Mice

Nakaaki Ohsawa

The nude mouse is a mutant with a congenital deficiency in the development of thymus and hair.[1] In order to determine whether the phenotypical expression of the *nu* gene is influenced or normalized by the coexistence of normal developmental tissues, the production of aggregation chimeras between nude mice and normal mice has been attempted.

MATERIALS AND METHODS

1. Animals used for chimera production
   For embryo aggregation, adult normal mice (C3H/HeN) of both sexes and adult nude mice (BALB/cA-*nu/nu*) of both sexes were used.
   As recipients, adult ICR-JCL mice of both sexes were used.

2. Culture medium
   Whittingham's medium[2] was used for the recovery of embryos, washing and *in vitro* culture of embryos through blastocysts.

3. Recovery of fertilized eggs
   Adult females of BALB/CA-*nu/nu* and C3H/HeN strains were injected with 5 IU of PMS (pregnant mare serum), followed by the injection of 5 IU of hCG (human chorionic gonadotropin) 48 hours later.

4. Aggregation chimeras[3]
   Recovered embryos of 8 to 16 cell stages were treated with Hanks solution containing 0.4% pronase at room temperature for 5 to 10 minutes to remove the zona pellucida. One embryo each of BALB/cA-*nu/nu* and C3H/HeN strains was placed in 20 to 30 μl of culture medium, and aggregation was performed by adding phytohemoagglutinin (PHA: 5μl/ml) to the standard medium. After the aggregation, the embryos stood in 5%

Third Department of Internal Medicine, University of Tokyo Faculty of Medicine, Tokyo, Japan

CO_2 and 95% air at room temperature for 15 minutes in order to complete the aggregation reaction. The aggregated embryos were cultured in 5% CO_2 and 95% air at 37°C for 24 to 40 hours, until the blastocysts developed. The chimeric blastocyst was about twice as big as a normal blastocyst. The chimeric embryos were transferred into the uteri of 2-day pseudopregnant recipient ICR female mice to yield littermates.

Obtained animals were analyzed for (1) Coat color and hair condition; (2) Thymus, by autopsy, and (3) Isozyme patterns of Es-3, Pep-3, Pgm-1 and Gpi-1 in thymus, brain, liver, and kidney by the electrophoretic method.

RESULTS AND DISCUSSION

Table 1 shows the summary of our results. Six hundred embryos each of BALB/cA-*nu*/*nu* and C3H/HeN strains were aggregated. Four hundred fifty-two aggregated embryos were obtained. Four hundred forty-six of these aggregated embryos were transferred to 32 recipient ICR female mice. Eighteen of the 32 recipients became pregnant and 52 pups were born. Of these, 39 pups grew up to be adult mice. The sex ratio of these 39 mice was female 25 (64%) and male 14 (36%).

TABLE 1.   Production of Chimeras between Nude Mice (BALB/c-*nu*/*nu*) and Normal Mice (C3H/HeN)

No. of embryo aggregated	600
No. of embryo transfered	446
No. of foster mothers	32
No. of successful pregnancies	18
No. of pups died postnatally	13
No. of pups died postnatully	13
No. of adult mice	39

Figure 1 shows an example of one litter. Three types of coat patterns were observed: C3H type, chimeric type and nude type.

Mice with the chimeric pattern coats were covered completely or incompletely with agouti and albino hairs. The results suggested that congenital deficiency of hair development due to the nu gene is corrected completely or partially by the influence of coexisting normal tissues during the developmental processes.

Table 2 summarizes the relationship between the status of coats and the

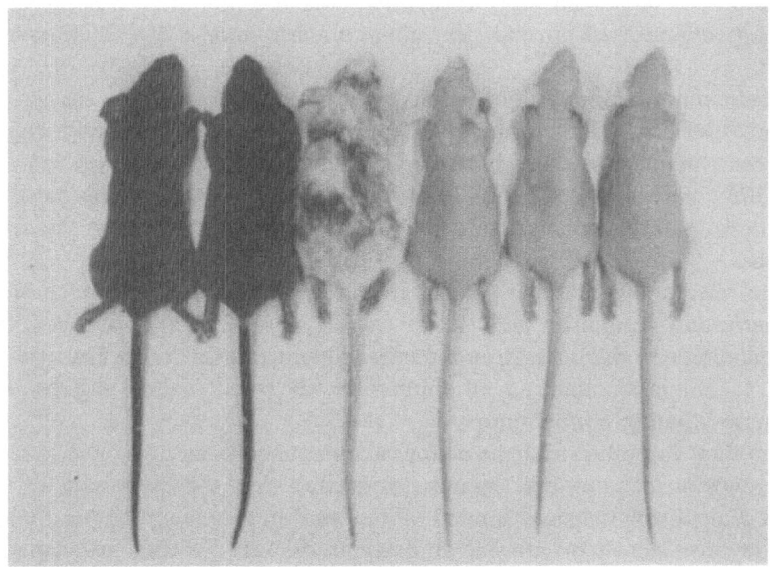

FIG. 1.   A litter of chimeras between nude mice (BALB/c-*nu*/ *nu*) and normal mice (C3H/HeN).

TABLE 2.   Morphological Characteristics of Nude ↔ Normal Chimeras

Coat color		C3H type	Chimeric type	Nude type
No. of mice		12	17	10
Thymus	+	12	1	0
	−	0	16	10
Hair	+	12	17 (15*)	0
	−	0	0	10

*partially depleted

thymuses in these 39 mice. The C3H type coat pattern occurred in 31% of the mice; the chimeric type, in 44%; and the nude type, in 21%.

Mice with C3H type coats were covered completely with agouti hairs, and mice with nude type coats were completely hairless. Out of 17 chimeric mice, two were completely covered with the chimeric hair and 15 showed partial hairlessness in the nude portion. All hairless or hair-poor regions were without pigment.

Mice with the C3H type coats showed thymuses of normal size. Mice with the nude type coats showed no thymus. Out of 17 chimeric type mice, 16 showed thymuses of normal size. One chimeric mouse was thymus-deficient.

Electrophoretic analysis of isozyme patterns in the thymus tissues revealed that all C3H type mice had only C3H type of Gpi-1 and Pep-3, whereas thymuses of chimeric type animals showed two types (C3H and BALB/c) of Pep-3, Gpi-1, Es-3 and Pgm-1. Analysis of isozyme patterns in other organs, such as kidney, liver and brain, revealed almost the same results.

The results suggested that the various organs in chimeric mice were chimeric and contained the cells of two strains, C3H and BALB/c. Although chimeric thymuses do not necessarily mean that the thymus epithelium is chimeric, analysis of chimerism of other organs suggested a chimeric thymus epithelium.

The fact that almost all the chimeric animals between nude and normal mice developed hairs and thymuses indicated that the expression of the nude gene during the development of hair and thymus might be suppressed by the coexistence of normal embryonal tissues. Further investigation is needed in order to clarify the above phenomena.

SUMMARY

Chimeras between nude and normal mice were successfully produced. These chimeras showed the chimeric coat patterns with hair and chimeric thymuses. The results suggest that the phenotypical expression of the *nu* gene is influenced by the coexisiting normal embryonal tissues.

Acknowledgements
The cooperation of Dr. Chikashi Tachi, Department of Zoology, University of Tokyo, and Mr. Minesuke Yokoyama, Central Institute for Experimental Animals, Kawasaki, is greatly appreciated.

REFERENCES

1. Rygaard, J.: Thymus and Self: Immunobiology of the Mouse Mutant Nude. F. A. D. L., Copenhagen, 1973.
2. Whittingham, D. G.: Culture of mouse ova. *J. Reprod. Fert.*, **Suppl. 14**: 7, 1971.
3. Mintz, B.: Allophenic mice of multi-embryo origin. In: Methods in Mammalian Embryology (ed. J. C. Daniel, Jr.), p. 186. Freeman & Co., San Francisco, 1971.

Discussion

Dr. Tachi: I think the understanding of the genetic factors which possibly regulate the processes of early embryogenesis is of primary importance for us to be able to use the experimental tools. We now have tools such as chimeras to use in experimentally analyzing the problems of cell lineage in mammalian embryos. Such knowledge will be important in answering the questions posed by the so-called "balanced" and "unbalanced" chimeras.

Search for Transplantation Method Facilitating Establishment of Normal Diploid Teratocarcinomas from "Resistant" C57BL Mice

Takehiko Noguchi,* Choji Taya*,** and Kazuo Moriwaki*

SUMMARY

The inguinal subcutaneous region has proven to be a much better transplantation site for teratocarcinomas than the lateral region which has been routinely used. Three transplantable teratocarcinomas were established from teratocarcinoma-resistant C57BL sublines and the B10.A congenic line. These teratocarcinomas were multipotent. G-banding karyotype analysis revealed that the modal karyotypes of these tumor lines were all normal diploid.

INTRODUCTION

Embryonal carcinoma cells, the stem cells of teratocarcinomas, are, in some cases, able to undergo completely normal differentiation when injected into blastocysts.[1] They are essentially totipotent and capable of differentiating not only into somatic tissues but also into functional germ cells.[1,2] Embryonal carcinoma cells can be mutagenized and screened in order to obtain cells bearing the desired genetic mutations. The mutant embryonal carcinoma cells, which have been normalized through the chimera-forming method, ultimately develop into mice bearing the mutations.[3]

Thus embryonal carcinoma cells have been attracting attention as vehicles for producing animal models of genetic diseases including nervous disorders.[3,4] In addition to developmental versatility, karyotypic normalcy would also be necessary.[5] Accordingly, normal diploid teratocarcinoma lines are needed as sources of these cell lines.

Murine teratocarcinomas can be experimentally induced by transplanting early stage embryos into the adult testis[6,7] or under the kidney

---

* National Institute of Genetics, Mishima, Shizuoka, Japan
** Medical School of Osaka University, Kita-ku, Osaka, Japan
Contribution No. 1375 from National Institute of Genetics

capsule.[8,9] This can be accomplished using animals from strains having high incidences of spontaneous congenital teratomas and from a few other inbred strains. However, strain differences in the inducibility and re-transplantability of primary teratocarcinomas are known.[6,10-12] Solter *et al.*[10,11] designated inbred strains with high yields and high retransplanta-bility of embryo-derived teratocarcinomas as "teratocarcinoma sensitive" or "permissive" strains, those with low yields and low transplantability were called "resistant" or "non-permissive" strains.

This paper describes a transplantation method which facilitated the establishment of multipotential normal diploid teratocarcinomas not only from sensitive strains but also from resistant strains.

Attempt to produce transplantable teratoma lines from a resistant strain, C57BL

Strain C57BL, which is one of the resistant strains,[10] is the most frequently used mouse strain. Subline C57BL/10Sn has been used as the inbred partner in developing many congenic resistant strains. Subline C-57BL/6 has been used to produce a wide range of mutants. If teratocarci-nomas can be obtained at will from these sublines, we can utilize many mutant and variant genes in the study of teratocarcinomas. We, there-fore, attempted to produce multipotential teratocarcinomas from these sublines of mice.

The genetic factor(s) responsible for strain differences seems to be mainly host factor(s).[10,11] The sensitive character of host mice is known to be genetically dominant.[10,11] Therefore, F_1 hybrids from a cross of C57BL and C3H (sensitive) were used as hosts for induction and transplantation of primary teratocarcinomas.

First we tried to induce teratocarcinomas from B10 H-2 congenic lines. Embryos at 6 1/2 days of gestation (the plug was found on day 0) were transplanted into the adult testis by the method described by Stevens.[6] Two teratocarcinomas, OTT10A-5 and OTT10Sn-3 were obtained from B10.A/SgSn and B10/Sn embryos, respectively.[13,14] After three subcu-taneous passages in the F_1 hybrids, the line OTT10A-5 was transplanted into syngenic hosts as well as F_1 hybrids. Unexpectedly, the tumor could be transplanted repeatedly in the resistant syngenic hosts. Futhermore, the transplantability of the tumor in the syngenic hosts was higher (86%, 50/58) than that of the F_1 hosts (56%, 29/55) for transplantation generations 3 to 6.

The transplantation behavior of OTT10Sn was similar to that of OTT10A. After two passages in the F_1 hosts, this tumor line could also be repeatedly transplanted in syngenic mice. In these two lines, transplantability in the syngenic hosts was not checked at the beginning of subcutaneous transplantation. Because of this, we investigated the possibility that the primary tumors might have been transplantable from the beginning.

Primary tumors were induced from 6 1/2 day-embryos from the C57BL/ 6 strain, which was reported to be a typical resistant strain.[11,12] Five primary teratocarcinomas were transplanted into the syngenic hosts, one of which (OKTB6-5) was repeatedly transplantable. The transplantation behaviors of these three lines in syngenic and in (C57BL × C3H) F_1 hosts are schematically presented in Fig. 1.

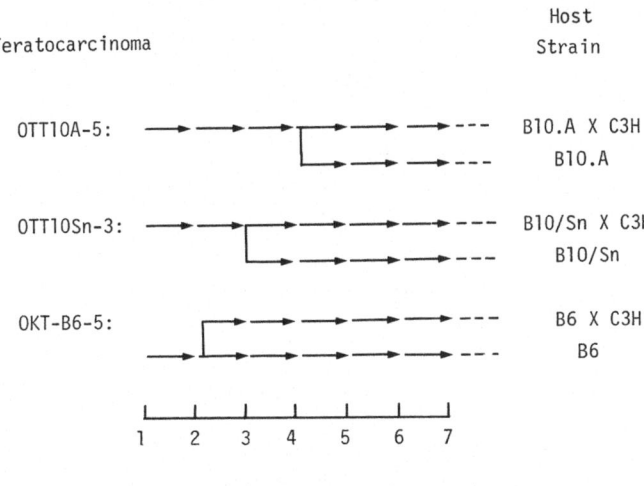

FIG. 1. Transplantation behavior of C57BL teratocarcinomas in syngenic and (C57BL × C3H) F_1 hosts.

Karyotype and developmental potency of C57BL teratocarcinomas

Modal karyotypes of these tumor lines were studied using G- and Q-banding techniques, and they were all found to be normal diploid. A typical normal diploid G-banded karyotype of a OTT10A cell is shown in Fig. 2.

The developmental potentials of these tumors were examined histologically. Solid tumors of these three lines contained derivatives from all three germ layers. The microphotos in Fig. 3 show histological sections of OTT-10A-5 solid tumors.

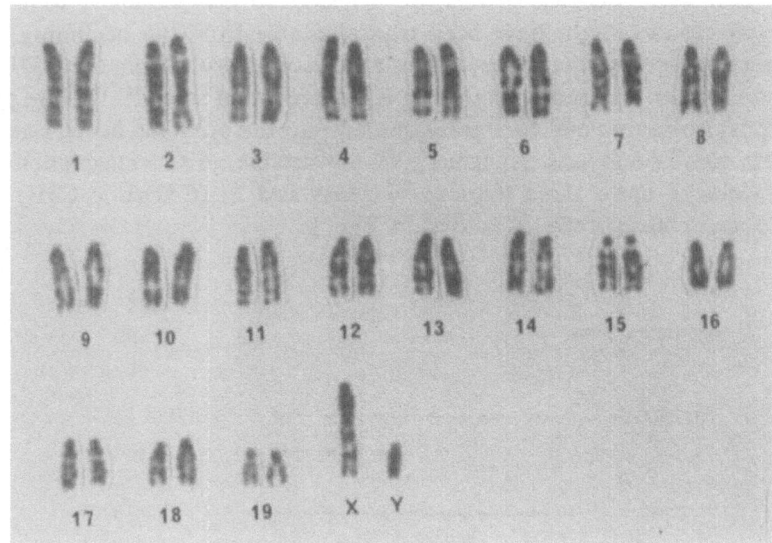

Fig. 2. A normal diploid G-banded karyotype (XY) obtained from OTT10A-5.

Growth of embryonal carcinoma cells in the lateral and the inguinal subcutaneous sites

The reason these multipotential and normal diploid teratocarcinomas could be established so easily from teratocarcinoma resistant mouse strains seemed to reside in the transplantation method that we used. The inoculation site widely used for teratocarcinoma transplantation is the lateral subcutaneous region (shaded area A in Fig. 4). But we used, instead, the inguinal site (shaded area B).

In this study, site B rather than site A was used because of observations made in preliminary experiments. In these experiments, we attempted to make embryoid bodies from a spontaneous testicular teratocarcinoma which occurred in a male of strain 129/Sv-*ter* having a high incidence of testicular teratomas.[15] The tumors used at that time were still primitive and had a high tendency to differentiate into benign tumors. Minced tumor tissue was injected into the intraperitoneal cavity. To prevent leakage of injected tumor tissues, trocars were first placed under the skin at the inguinal site, inserted near the center of the abdomen through the subcutaneous

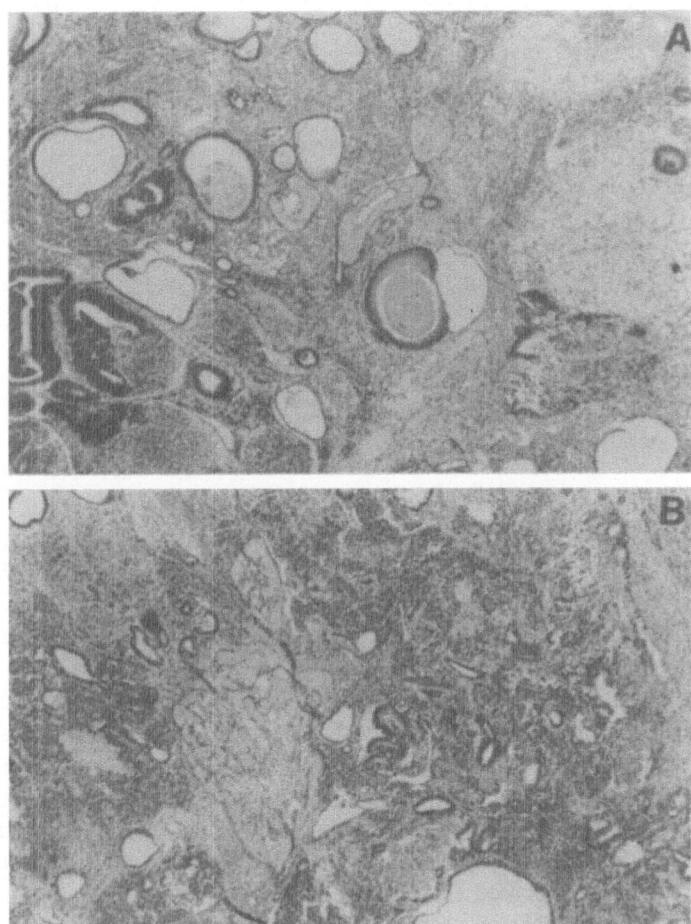

Fig. 3. Microphotographs of sections of OTT10A-5 solid tumors produced by subcutaneous transplantation.

A: A tumor from the 2nd subcutaneous transplantation generation. B: A tumor from the 5th generation. In these tumors, neural tissues, muscles, keratin parles, adipose tissues, bones, melanocytes, endodermal cysts, mesenchymes, embryonal carcinoma cells, etc. can be seen.

T. Noguchi *et al.*

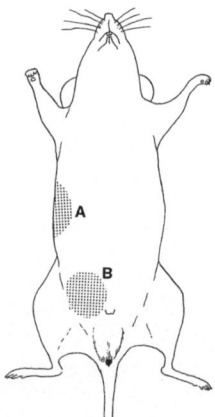

FIG. 4. Subcutaneous transplantation sites. Shaded area
A: lateral region. Shaded area B: inguinal region.

space, and into the muscle layer into the intraperitoneal cavity. Most of the tissue was left in the cavity. Tumors were, however, formed more often near the inguinal subcutaneous site than in the intraperitoneal cavity. This was often observed with other primitive teratocarcinomas. Thus we thought that the inguinal site might be a preferable transplantation site for teratocarcinomas.

To investigate this further, defined doses of embryonal carcinoma cells from primitive tumors of OTT10A and OTT10Sn (from the 2nd subcutaneous transplantation generations) were injected into site A and site B simultaneously, and the number of tumors formed at these two sites was compared. Hosts animals were autopsied after 4 weeks for OTT10A, and after 3 weeks for OTT10Sn. The results are shown in Table 1. It is clear that far more tumors were formed at site B than at site A. Figure 5 shows some of the B10.A hosts injected with OTT10A embryonal carcinoma cells. The tumors were all histologically identified as teratocarcinomas.

It may be concluded that the inguinal subcutaneous region is a better transplantation site for teratocarcinomas than the lateral region.

DISCUSSION

Difficulty in producing transplantable teratocarcinomas from resistant strains seems to be mainly due to the poor growth capacity of the stem cells of primitive teratocarcinomas in the subcutaneous site. The subcutaneous site that has been routinely used is the lateral region. Even in sensitive strains,

TABLE 1.   Growth of Embryonal Carcinoma Cells at Two Subcutaneous Transplantation Sites

Teratocarcinoma	Inoculum* size /site	No. of mice injected	No. of mice with tumors**		
			at site A	at site B	at sites A & B
OTT10A-5	6×10^5	20	0	8	0
OTT10Sn-3	3×10^5	16	0	11	1

* Embryonal carcinoma cells grown *in vitro* were used. Enrichment of the stem cells of these teratocarcinomas was attained by passaging several times at 2 to 3-day intervals on feeder layer (mitomycin C-treated 3T3-A31 fibroblasts) in Dulbecco's MEM-10% FCS medium. In the last passage cells were grown on gelatin-coated dishes in the medium supplemented with 10^{-4} M β-mercaptoethanol[16] to diminish the relative fraction of 3T3 cells. Embryonal carcinoma cells were distinguished by size from 3T3 cells under an inverted microscope, and counted using a hematocytometer.
** Mice were autopsied at 4 weeks in OTT10A-5 and 3 weeks in OTT10Sn-3. Visible tumors were examined histologically and were identified as teratocarcinomas.

FIG. 5.   B10.A mice injected with embryonal carcinoma cells from OTT10A-5 tumor at the 2nd transplantation age. Arrows show the tumors formed at the inguinal site. None of them had tumors at the lateral site.

primary teratocarcinomas have, in general, a low retransplantability at the lateral subcutaneous region. This may be due to the tendency of the primitive stem cells to differentiate and to the low number of stem cells in the primary tumors.

The experimental results presented here show that the inguinal subcutaneous site is a much better transplantation site for primitive terato-carcinoma cells than those sites commonly used. This transplantation site supported the growth of embryonal carcinoma cells of tumors from resistant strains as well as tumors from sensitive strains. Teratocarcinomas maintained at this site had mostly normal diploid modal karyotype, at least within several transplantation generations. Normal diploid teratocarcinomas can probably be obtained from most inbred strains of the mouse since C57BL is a typical teratocarcinoma resistant strain.[10,11] If normal diploid teratocarcinomas can be produced at will from any strain of mouse, the variety of genetic backgrounds of teratocarcinomas will be greatly increased. Hence, the utility of teratocarcinomas as vehicles for the introduction of variant or mutant genes into live mice will also be expanded.

A question raised here may be whether the strain difference in retransplantability of primary teratocarcinomas observed at the lateral site is nullified at the inguinal site. Table 2 lists the teratocarcinomas that were successfully transplanted to the inguinal site. The numbers attached to each tumor line indicates the serial number of primary teratocarcinomas from which transplantable lines were obtained. The numbers show that larger numbers of trials were needed for producing transplantable lines in

TABLE 2.   Multipotential Teratocarcinomas Established by the New Transplantation Method

Origin (mouse)	Teratocarcinoma	Derivation	Modal karyotype	Tumor age at karyotype preparation
Resistant strain				
B10.A/SgSn	OTT10A-5	6-day embryo grafted in the testis	normal diploid (XY) in G-band pattern	8 months
C57BL/10Sn	OTT10Sn-3	6-day embryo grafted in the testis	normal diploid (XY) in G-band pattern	5 months
C57BL/6J	OKTB6-5	6-day embryo grafted in the kidney capsule	normal diploid (XX) in G-band pattern	4 months
Sensitive strain				
129/Sv-*ter*	STT-2	germ cell, spontaneous testicular teratoma	normal diploid (XY) abnormal (S=40:M2) in conventional staining	5 months 8 months
129/Sv-*Sl*-CP	OKT129-1	6-day embryo grafted in the kidney capsule	not checked	——
C3H/HeJ	OKTC3H-1	6-day embryo grafted in the kidney capsule	normal diploid (XY) in conventional staining	2 months

C57BL mice than in sensitive 129 and C3H mice. Thus, strain differences in transplantability still seem to exist at the inguinal site. But the difficulty in obtaining transplantable lines from teratocarcinoma-resistant strains will be lightened by utilizing the inguinal transplantation site.

Acknowledgements

The authors would like to thank Dr. Leroy C. Stevens and Mr. Don S. Varnum of the Jackson Laboratory for teaching the technique of experimental induction of embryo-derived teratocarcinomas.

This research was supported in part by Grants-in-Aid for scientific research from the Ministry of Education, Science and Culture.

REFERENCES

1. Mintz, B. and Illmensee, K.: Normal genetically mosaic mice produced from malignant teratocarcinoma cells. *Proc. Nat. Acad. Sci. U.S.A.*, **72**: 3585–3589, 1975.
2. Illmensee, K. and Mintz, B.: Totipotency and normal differentiation of single teratocarcinoma cells cloned by injection into blastocysts. *Proc. Nat. Acad. Sci. U.S.A.*, **73**: 549–553, 1976.
3. Dewey, M. J., Martin, D. W., Jr., Martin, G. R. and Mintz, B.: Mosaic mice with teratocarcinoma-derived mutant cells deficient in hypoxanthine phosphoribosyltransferase. *Proc. Nat. Acad. Sci. U.S.A.*, **74**: 5564–5568, 1977.
4. Mintz, B.: Teratocarcinoma cells as vehicles for introducing mutant genes into mice. *Differentiation*, **13**: 25–27, 1979.
5. Cronmiller, C. and Mintz, B.: Karyotypic normalcy and quasi-normalcy of developmentally totipotent mouse teratocarcinoma cells. *Develop. Biol.*, **67**: 465–477, 1978.
6. Stevens, L. C.: The development of teratomas from intratesticular grafts of tubal mouse eggs. *J. Embryol. exp. Morph.*, **20**: 329–341, 1968.
7. Stevens, L. C.: The development of transplantable teratocarcinomas from intratesticular grafts of pre- and post-implantation mouse embryos. *Develop. Biol.*, **21**: 364–382, 1970.
8. Solter, D., Skreb, N. and Damjanov, I.: Extrauterine growth of mouse egg-cylinder results in malignant teratoma. *Nature*, **227**: 503–504, 1970.
9. Illes, S. A.: Mouse teratomas and embryoid bodies: Their induction and differentiation. *J. Embryol. exp. Morph.*, **38**: 63–75, 1977.
10. Solter, D., Adams, N., Damjanov, I. and Koprowski, H.: Control of teratocarcinogenesis. In: Teratomas and Differentiation (ed. M. I. Sherman and D. Solter), pp. 139–159. Academic Press, Inc., New York, 1975.
11. Solter, D., Damjanov, I. and Koprowski, H.: Embryo-derived teratoma: A model system in developmental and tumor biology. In: The Early Devel-

opment of Mammals (ed. M. Balls and A. E. Wild), pp. 243–264. Cambridge University Press, Cambridge, 1975.

12. Solter, D., Dominis, M. and Damjanov, I.: Embryo-derived teratocarcinoma: I. The role of strain and gender in the control of teratocarcinogenesis. *Int. J. Cancer,* **24**: 770–772, 1979.

13. Noguchi, T., Taya, C. and Moriwaki, K.: A normal diploid teratocarcinoma (OTT10A-5) obtained from a 6 day embryo of B10.A/SgSn mouse (Abstract). *Develop. Growth & Differ.,* **22**: 715, 1980.

14. Noguchi, T., Taya, C., Shiroishi, T., Noguchi, M., Nishimune, Y., Ogiso, Y., Matsushiro, A. and Moriwaki, K.: Studies on normal diploid teratocarcinomas derived from B10 H-2 congenic mice. In: The Proceedings of the 1st Hiei International Symposium on Teratomas and Cell Surface (ed. T. Muramatsu and Y. Ikawa). Japan Sci. Soc. Press, Tokyo (in press).

15. Stevens, L. C.: A new inbred sublines of mice (129/*ter*Sv) with a high incidence of spontaneous congenital testicular teratomas. *J. Nat. Cancer Inst.,* **50**: 235–242, 1973.

16. Oshima, R.: Stimulation of the growth and differentiation of feeder layer dependent mouse embryonal carcinoma cells by β-mercaptoethanol. *Differentiation,* **11**: 149–155, 1978.

Discussion

Dr. Stevens: 1. When you inject the teratocarcinomas abdominally, you get both subcutaneous and intraperitoneal tumors. Are they histologically different?

     2. Do you get embryoid bodies?

Dr. Noguchi: 1. I do not have enough data to be able to answer your question, but I got the impression that subcutaneous tumors seemed to contain more differentiated tissue than intraperitoneal tumors.

     2. I have not gotten embryoid bodies yet from C57BL teratocarcinomas, but I got them from a testicular teratocarcinoma.

The Use of Chimeric Rats in the Analysis of the Hooded Pigmentation Pattern

Ken-ichi Yamamura,* Zen-ichi Ogita,** and Clement L. Markert***

Chimeric mice have been produced on a large scale in several laboratories by techniques of embryonic manipulation. Two or more embryos may be aggregated during early cleavage stages to produce a single embryo,[1-3] or cells from one embryo may be injected into the blastocyst cavity of another, to become incorporated into the inner cell mass and thus produce a chimeric embryo.[4] Such techniques have been applied to other mammals, particularly rabbits,[5,6] sheep,[7] and rats,[8,9] though with only limited success. These chimeras have been used in the studies of mammalian development.[10,11] A total of four chimeric rats have so far been reported in the literature.[8,9] We have improved the techniques for producing chimeric rats and a total of twenty-seven have been born and raised to maturity. These chimeric rats have proved useful in the analysis of the hooded pigmentation pattern of rats.

PRODUCTION OF CHIMERIC RATS

The rats were maintained on a 14-hour light, 10-hour dark schedule (lights on at 5 AM, off at 7 PM).

The rat strains used in this investigation were derived from Sprague-Dawley rats (CD, Charles River, Wilmington, Massachusetts), that is, albino hooded, and from a BD IX strain maintained in our laboratory. Matings between these two strains led to the extraction of a black hooded strain, an agouti hooded strain, and an albino non-hooded strain.

The most suitable time for hormonal injections on any day is between 4 and 5 PM. Five IUs of PMSG (pregnant mare's serum gonadotropin) and five IUs of HCG (human chorionic gonadotropin), commonly used with mice, are not enough to cause superovulation in rats. Twenty IUs

* Department of Medicine and Geriatrics, Osaka University Medical School, Osaka, Japan
** Department of Biochemical Pathology, Research Institute for Wakan-Yaku, Toyama Medical and Pharmaceutical University, Toyama, Japan
*** Department of Biology, Yale University, New Haven, Counecticut, U.S.A.

of each hormonal preparation is too much, and less effective. Ten IUs of each proved to be the most effective.

We injected hormones at various stages of the estrous cycle in order to examine the effect of the estrous cycle on superovulation. As shown in Tables 1 and 2, we found that it was very important to inject the PMS at the metestrous stage of the estrous cycle, followed 48 hours later by an injection of HCG. Injections out of phase with the estrous cycle may in fact induce superovulation, but the eggs are abnormal and the resulting embryos do not survive far more than two or three cleavages. The requirement of synchronization of hormonal injections with the estrous cycle makes physiological sense, but indicates that the rat is more restricted in its responses than is the mouse.

TABLE 1.   Number of Eight-cell Stage Embryos Obtained from 8 to 13-week-old Female Rats

	Spontaneous ovulation	Superovulation	
	Average (embryos/rats)	Not synchronized	Synchronized
BD IX	7.3 (154/21)	4.3 (43/10)	n.d.
Black	10.4 (290/28)	6.9 (76/11)	20.1 (302/15)
Agouti hooded	7.6 (76/10)	n.d.	13.5 (27/2)
Black hooded	8.0 (40/5)	7.0 (7/1)	8.2 (41/5)
CD	7.3 (87/12)	12.0 (36/3)	19.0 (38/2)
Total	8.2 (648/79)	6.5 (162/25)	17.0 (408/24)

n.d. = not done

TABLE 2.   Effect of the Estrous Cycle on the Number of Embryos Superovulated from 8 to 13-week-old Female Rats

Estrous stage at PMSG injection	Number of rats tested	Number of embryos	Average
proestrous	3	0	0.0
estrous	4	38	9.5
metestrous	6	130	21.7
diestrous	4	14	3.5

The mating rate is also increased by this synchronization of hormone injection (Table 3). Thus, three to four times as many eight-cell-stage embryos can be obtained from female rats superovulated by administrating 10 IU of PMS in 0.1 ml of normal saline by intraperitoneal injection at the

TABLE 3.   Mating Rate

	Spontaneous ovulation (within 5 days)	Superovulation	
		Not synchronized	Synchronized
BD IX	47.9% (23/48)	66.7% (10/15)	n.d.
Black	67.9% (38/56)	75.0% (12/16)	93.8% (15/16)
Agouti hooded	50.0% (11/22)	n.d.	100.0% (2/2)
Black hooded	40.6% (13/32)	50.0% (1/2)	83.3% (5/6)
CD	44.4% (12/27)	87.5% (7/8)	100.0% (2/2)
Total	52.4% (97/185)	73.2% (30/41)	92.3% (24/26)

n.d. = not done

metestrous stage of the estrous cycle, followed 48 hours later by an injection of 10 IU of HCG as from females who ovulated spontaneously.

Culture of eight-cell-stage embryo

Although rat embryos can develop from the eight-cell to the blastocyst stage in ordinary mouse medium, the success rate is about 50% (unpublished data). Mayer and Fritz[8] cultured eight-cell rat embryos with about 80% success. When late eight-cell embryos were cultured in the modified culture medium[12] immediately after flushing from the oviduct, about 90% of them developed into normal blastocysts. However, only about 80% developed normally when they were prepared for aggregation as chimeras. Thus, any manipulation of the embryos, such as removing the zona pellucida or treating them with phytohemaglutinin, damages some embryos. Older embryos survived better in the culture medium, but for aggregation purposes, eight-cell-stage embryos must be used. In order to solve this problem, we usually killed rats on the afternoon of day 3, by which time most of the rat embryos had developed into the late eight-cell-stage. Another important concern is that the embryo aggregation should be done as quickly as possible. We usually aggregated about 20 pairs of embryos in one experiment. Clearly, the culture media are marginal at best, and even modification and supplementation only slightly improved the survival and growth of the embryos in culture.

Aggregation of embryos

As pointed out by Mayer and Fritz,[8] rat embryos do not aggregate well. One reason is that most rat embryos are disk-shaped rather than globular in shape like mouse embryos. Thus, if a pair of embryos are merely pushed together and attach along one edge, they do not develop into a single, integrated blastocyst. Instead, two attached embryos develop. We found that by aggregating two embryos so that the large flat surfaces were apposed (Fig. 1), about 50% of the pairs developed into single, integrated morulae or blastocysts without the use of any special culture dish such as that used

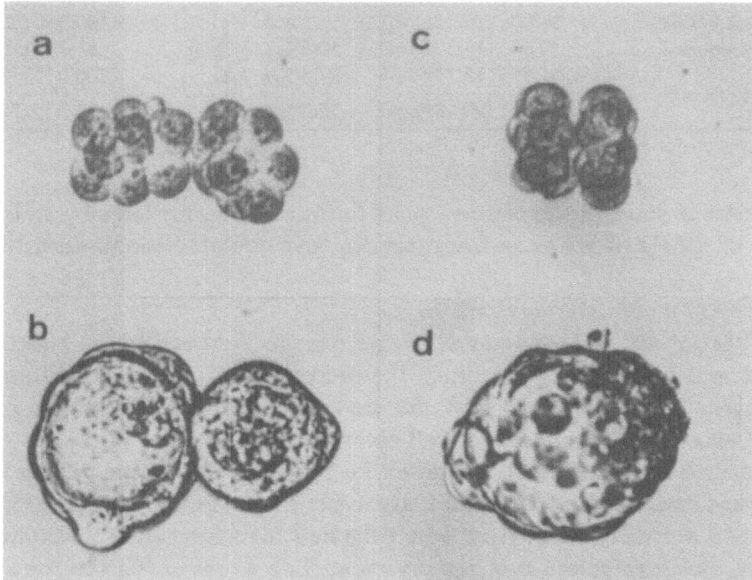

FIG. 1.   Aggregation of two embryos. If a pair of embryos were merely pushed together and allowed to attach along one edge (a), then these embryos did not develop into a single, integrated blastcyst. Instead, two attached embryos develop (b). If a pair of embryos were aggregated with the flat surfaces apposed (c), then they developed into a single blastocyst (d).

by Mayer and Fritz.[8] As described above, 80% of the treated embryos will develop into blastocysts, we anticipated that about 64% of the chimeric aggregates would develop into single blastocysts, but our success rate was lower. About 14% of our chimeric aggregates separated and failed to develop into single, integrated blastocysts.

Production and coat colors of chimeric rats

In a series of experiments, a total of 233 aggregated pairs of embryos were transferred to recipient females and 27 chimeric rats were born. Five black ↔ black hooded, six black ↔ albino hooded (CD), five black ↔ agouti hooded, ten black ↔ albino non-hooded, and one agouti (BD IX) ↔ pink-eyed dilute hooded (RCS) chimeras were produced from 58, 44, 78, 39, and 10 aggregates, respectively.

The coat color patterns of black ↔ agouti hooded chimeras displayed both agouti and black pigmentation over the entire surface of the chimera. Black ↔ black hooded chimeras displayed a white belly spot. In black ↔ albino hooded chimeras, small patches of white hair appeared on the head and a white spot occurred on the belly. In an agouti ↔ pink-eyed dilute hooded chimera, the pigment cells from the hooded pink-eyed dilute component were only expressed in a small area of the head.

ANALYSIS OF THE HOODED PIGMENTATION PATTERN

Rats of the hooded phenotype have been known since the last century. Breeding tests early revealed that this character is inherited as a simple monogenic recessive. The expression of the hooded pattern is relatively stable in comparison with the erratic spotting of the ordinary piebald mouse. So far, three recessive alleles of the hooded locus have been discovered: h, h^i, and h^n. The proportion of pigmented hair in each homozygous phenotypes of the three alleles (h/h, h^i/h^i, h^n/h^n) is about 35%, 83%, and 10%, respectively.[13] Hooded rats in our colony have a black head and shoulders with a sometimes interrupted black stripe down the center of the back to the tail. The rest of the body is white with an occasional black spot in the white area.

Unpigmented hair in potentially pigmented rats can be caused by abnormalities at any step of melanoblast differentiation or in the differentiation of the tissue environment, primarily the cells of the hair follicles. To explore this problem, we have produced genetically single-labeled and double-labeled chimeric rats with alternate allels at the C and A loci.

Causes of unpigmented hair in hooded rats

There is direct evidence that pigment cells are absent from the white areas of the hooded rat.[14] Our own histological studies on the white hairs of hooded rats demonstrated no large clear cells (amelanotic melanocytes) in the upper portion of the matrix of the hair bulbs.[12] Such large clear cells can be found in hairs in the hooded areas of albino hooded rats and in non-hooded albinos. Moreover, pigmented hairs can be formed in the prospective white-haired ventral skin if it is transplanted to a pigmented region at birth. In addition, pigment hairs can arise from the prospective white ventral skin that lies adjacent to the pigmented graft.[15] These results suggest that the skin of the white ventral area is normally free of pigment cells.

In our experiments, clear cells were not found in hairs from the white belly spots of the chimeric rats. On the other hand, clear cells were found in hairs from the white area of the head of black ↔ albino hooded chimeras. Thus, we conclude that the prospective white haired area of the hooded rat is free of pigment cells as in the white belly spot of the hooded ↔ non-hooded chimera.

Site of hooded gene expression

The absence of pigment cells in hair follicles can be attributed to the blockage of the differentiation of melanoblasts, and/or to the nature of the cellular environment (particularly the hair follicles).

If the hooded alleles, in the cells of the hair follicles, acted to prevent melanoblasts from entering, then black ↔ agouti hooded chimeras should show a diffuse distribution of black and white, or stripes, in the prospective white haired area (Fig. 2) instead of the black and agouti pigmentation

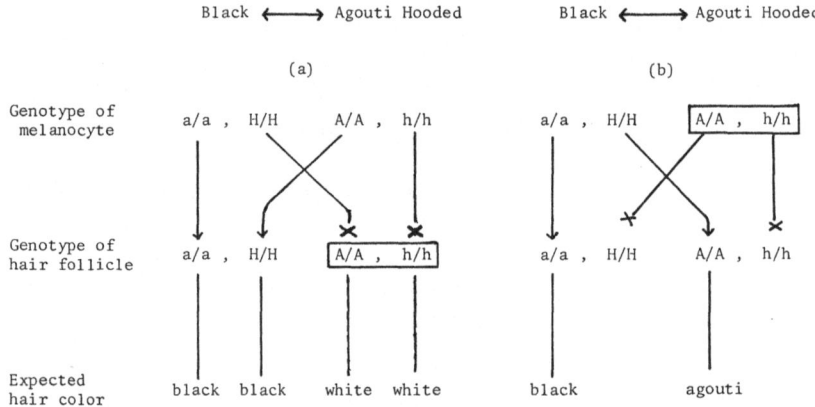

FIG. 2. Four possible combinations of melanoblasts and hair follicles in black ↔ agouti hooded chimeras. If the hair follicle of the hooded genotype prevented the melanoblast from entering, then these chimeras should show mixed areas of black and white hairs (a). If the melanoblasts of the hooded genotype were not differentiated for some reason, these chimeras should show areas of both black and agouti pigmentation (b). As described in the text, both black and agouti hairs were seen in these chimeras.

actually observed throughout the coat. We observed that the skin of the hooded genotype readily formed hair follicles that produced pigmented hairs of the agouti phenotype in black $\leftrightarrow$ agouti hooded chimeras. As previously mentioned, the experiments done by Rawles[15] showed that the pigment cells migrate into the hair follicles of the prospective white-haired ventral skin. Both observations suggest that the hair follicles of the hooded genetic make-up do not prevent melanoblasts from entering and maturing into melanocytes. Thus, the hooded gene may be acting within the melanoblast itself and not within the cells making up the environment of the pigment cell.

Effect of the hooded gene on the melanoblast

If the hooded gene is acting within the melanocyte, then the absence of pigment cells in the hair follicles from the white areas could result from: 1) failure of melanoblasts to survive after penetration into the hair follicles; 2) failure of melanoblasts to penetrate the hair follicles; 3) failure of melanoblasts to multiply and migrate; or 4) failure of melanoblasts to differentiate in the posterior regions of the neural crest.

Once the melanoblast population of a hair follicle is destroyed, all subsequent generations of hairs produced by that follicle remain unpigmented, although the hairs are normal in every structural detail.[16] Consequently, if a hooded melanoblast entered but failed to survive in the flank and ventral hair follicles of the hooded $\leftrightarrow$ non-hooded chimera, then the flank and ventral fur of these chimeras should show areas of both pigmented and white hairs instead of the belly spot actually observed. Thus the first hypothesis seems an unlikely explanation for the white belly spot of chimeras or for the white areas of the hooded phenotype.

A diluted pigment pattern with scattered white hairs can be produced by irradiation that kills a proportion of the melanoblasts.[17] Such treatment can simulate the consequences of genetic death of melanoblasts due to the failure of penetration into the hair follicles. However, this result was not observed. Moreover, if the melanoblasts failed to penetrate the hair follicles after migration, then the hooded $\leftrightarrow$ non-hooded chimeras should show a diluted pigment pattern with scattered white hairs not only in the flank or belly but in the dorsal region. No such pigmentation pattern was ever observed. Consequently, the second hypothesis is also an unlikely explanation.

Developmental defects in the neural crest of the posterior region of hooded animals is an another possible explanation for the white pattern. However, all identified derivatives of neural crest cells appeared normal. Pigment cells do arise along the entire length of the neural crest since there is commonly a dorsal band of pigmentation extending to the tail. These observations seem to eliminate the fourth hypothesis.

Our data seem to support the third hypothesis. Wendt-Wagener[18] demonstrated a one-day delay in the migration of hooded melanoblasts in the head region, and at least a two-day delay in melanoblast migration in the posterior trunk region of the hooded rat. Apparently melanoblasts fail to reach the epidermis of the prospective white-haired areas because of the increasing density of the connective tissue as development proceeds. In black ↔ albino hooded chimeras, no white spot was ever found along the mid-line on the back, though white areas were found on the dorsal part of the head. This means that in the head regions, melanoblasts of the hooded genotype can compete effectively with other melanoblasts, but fail to do so in the posterior regions of the body. This implies that the migration speed of melanoblasts carrying the hooded alleles is slower in the posterior regions of the body.

Then, why can melanoblasts of the hooded genotype in the head region compete effectively with other melanoblasts? This can perhaps best be explained as follows. 1) The neural crest first starts to develop and migrate in the mid-brain region of the developing embryo.[19] Melanoblasts differentiating in this area may have sufficient time to migrate to the hair follicles in the head and shoulder region. 2) It is known that the neural crest is considerably more massive in the head region than in the trunk region in chick embryos,[20] and this may be true in rats. 3) The head and trunk neural crests differ in their developmental capacities. In contrast to the mesodermally-derived mesenchyme of the trunk, the head mesenchyme is substantially derived from the neural crest. Thus, the early mesenchymal tissue environment confronting migrating melanoblasts changes along the anterior-posterior axis of the embryo.

The formation of the white belly spot of the hooded ↔ non-hooded chimeras can be explained as follows. In these chimeras, in the posterior regions of the body, the melanoblasts carrying the hooded alleles seem to migrate slower than do normal melanoblasts. The normal melanoblasts, on the other hand, can migrate all the way to the mid ventral line. However, if the normal melanoblasts are only a small percentage of the melanoblast population that migrates from the neural crest, then they may be able to populate only the follicles of the dorsal regions, leaving the ventral hair follicles unpopulated causing the white belly spot. The variability in the size of the white belly spot in our chimeras can be explained by assuming different proportions of normal and hooded melanoblasts, i.e., a greater proportion of hooded melanoblasts results in a larger white belly spot.

Acknowledgements
This study was supported by NIH grant 5 ROI HD 07741 and by grant #NSF DAR 7910054. KY was supported by a Rockefeller Foundation Postdoctoral Fellowship.

REFERENCES

1. Tarkowski, A. K.: Mouse chimeras developed from fused eggs. *Nature*, **190**: 857–860, 1961.
2. Mintz, B.: Formation of genotypically mosaic mouse embryos. *Amer. Zool.*, **2**: 432, 1962.
3. Markert, C. L. and Petters, R. M.: Manufactured hexaparental mice show that adults are derived from three embryonic cells. *Science*, **202**: 56–58, 1968.
4. Gardner, R. L.: Mouse chimeras obtained by the injection of cells into the blastocyst. *Nature*, **220**: 596–597, 1968.
5. Gardner, R. L. and Munro, A. J.: Successful construction of chimeric rabbit. *Nature*, **250**: 146–147, 1974.
6. Moustafa L. A.: Chimeric rabbits from embryonic cell transplantation. *Proc. Soc. Exp. Biol. Med.*, **147**: 485–488, 1974.
7. Tucker, E. M., Moor, R. M. and Rowson, L. E. A.: Tetraparental sheep chimeras induced by blastomere transplantation. *Immunology*, **26**: 613–621, 1974.
8. Mayer, J. F. and Fritz, H. I.: The culture of preimplantation rat embryos and the production of allophenic rats. *J. Reprod. Fert.*, **39**: 1–9, 1974.
9. Mullen, R. J. and Lavail, M. M.: Inherited retinal dystrophy: Primary defect in pigment epithelium determined with experimental rat chimeras. *Science*, **192**: 799–801, 1976.
10. Mclauren, A.: Mammalian Chimeras. Cambridge University Press, London, 1976.
11. Russell, L. B. (ed.): Genetic Mosaics and Chimeras in Mammals. Plenum Press, New York, 1978.
12. Yamamura, K. and Markert, C. L.: The production of chimeric rats and their use in the analysis of the hooded pigmentation pattern. *Dev. Genet.*, **2**: 131–146, 1981.
13. Robinson, R.: Genetics of the Norway Rat-Color Variation. Pergamon Press, London, 1965.
14. Taylor, A. G.: Survival of rat skin and changes in hair pigmentation following freezing. *J. Exp. Zool.*, **110**: 77–112, 1949.
15. Rawles, M. E.: Pigmentation in autoplastic and homoplastic grafts of skin from fetal and newborn hooded rats. *Amer. J. Anat.*, **97**: 79–128, 1955.
16. Rawles M. E.: Origin of the mammalian pigment cell and its role in the pigmentation hair. In: Pigment Cell Growth (ed. M. G. Gordon), pp. 1–15. Academic Press, Inc., New York, 1953.
17. Chase, H. B.: Number of entities inactivated by X-rays in greying of hair. *Science*, **113**: 714–716, 1951.

18. Wendt-Wagener, G.: Untersuchungen über die Ausbreitung der Melano-
blasten bei einfarbig schwarzen Ratten und bei hauben Ratten. *Z. Verebungsl.,*
92: 63–68, 1961.
19. Morriss, G. M. and Thorogood, P. V.: An approach to cranial neural crest
cell migration and differentiation in mammalian embryos. In: Development in Mammals (ed. M. J. Johnson), pp. 363–412. North-Holland,
Amsterdam, 1978.
20. Romonoff, A. L.: The Avian Embryo: Structural and Functional Development, pp. 222–225. Macmillan Company, New York, 1960.

Quail-Chick Chimeras, a Tool to Study the Development of the Peripheral Nervous System

Nicole M. Le Douarin

INTRODUCTION

In the genetic approach to problems of developmental biology, one of the most productive methods has been to divise embryonic chimeras. Combination of embryos of different strains of mouse[1,2] has been and still is a fertile avenue for research on the development of mammals (see McLaren, 1976).[3] Thanks to its availability in the egg during the whole period of incubation, the avian embryo allows very precise types of chimeras to be constructed by transplanting into the embryo-defined tissular areas at definite developmental stages. This convenience has been made particularly profitable when tissues of two species of birds, whose cells can be easily differentiated, have been combined. In the quail and chick species, the nuclei differ during the interphase by the presence in quail cells of a large mass of heterochromatin associated with the nucleolus. This provides a stable marker which makes quail and chick cells readily identifiable at the light and electron microscopic levels.[4,5] The quail-chick marker system has been extensively used in my laboratory and in others to study the development of the neural crest.[6-9]

The neural crest is a transitory structure of the vertebrate embryo from which practically all the cell components of the peripheral nervous system (PNS) are derived. Neural crest cells undergo precise and extensive migration in the embryo and then settle in various embryonic primordia where they give rise to a number of cell types (see reference 7 for a review).

The quail-chick chimera system was first used to study the migration pattern of neural crest cells and to determine the origin in the neural primordium of the various structures belonging to the PNS.

1. Origin of peripheral ganglia from the neural crest

The experimental procedure involved replacement of regions of the neural primordium (neural tube together with neural folds) in a chick host

Institut d'Embryologie du CNRS et du Collège de France, Nogent-sur-Marne, France by the equivalent portion from a quail donor at the same developmental stage or vice versa.[10]

The results of these experiments showed unequivocally that the migration of cervico-dorsal neural crest cells from the region between somites 7 and 28 is confined to the dorsal mesenchyme derived from the somites and the intermediate cell mass. Neural crest derivatives in this area (except for melanocytes and Schwann cells) were found to be restricted to the sensory and sympathetic chain ganglia, the aortic and adrenal plexuses, and the adrenomedullary paraganglia. In other words, ganglion cell precursors from the trunk region do not penetrate the dorsal mesentery. In contrast, neural crest cells from the vagal (somites 1–7) and lumbosacral (behind somite 28) levels invade both the dorsal mesenchyme and the splanchnopleure. The myenteric plexuses are derived primarily from the crest at the level of somites 1–7; the lumbosacral crest contributes only to the ganglion of Remak and to some ganglia of the postumbilical gut.

The neural crest can therefore be divided into several different axial regions with respect to the development of the autonomic nervous system. One (level of somites 1–7) gives rise to the enteric ganglia of the gut. Another (somites 7–28) is the precursor of sympathoblasts of the trunk (within this area, crest at the level of somites 18–24 contributes to the adrenal medulla), and, finally, there are two regions from which both enteric and sympathetic ganglion cells arise (somites 5–7 and behind the 28th pair of somites).

2. Lability of the differentiation options of neural crest derived cells

Thus, although seemingly homogeneous, the crest cell population appears to be heterogeneous with respect to its migration behavior and its potentialities to differentiate into sympathetic and parasympathetic ganglia (i.e., to synthesize catecholamines (CA) or acetylcholine as neurotransmitters). Is this early determination a stable one or can the fate of neural crest cells be changed by artificially altering their initial localization along the axis of the embryo?

Experiments were done to find out whether the migratory behavior and phenotypic expression of crest cells from different axial levels were irreversibly fixed before migration and if experimental transposition of the axial levels would lead to disturbances in the development of crest derivatives. Neural primordia were transplanted heterotopically between quail and chick embryos. Histological examination of the resulting chimeras showed that crest cells of the adrenomedullary region grafted at the vagal level were able to colonize the gut (which they never do in normal development) and give rise to functional cholinergic ganglia. Conversely, crest cells from the mesencephalon or rhombencephalon, grafted at the adrenomedullary level, produced adrenergic sympathetic ganglia, populated the adrenal gland, and differentiated into adrenomedullary-like cells.[11,12]

These results indicate that the migratory behavior of crest cells depends on the pathways available when they leave the neural primordium rather than on some specificity related to their origin in the neural axis. The vagal and the adrenomedullary regions of the embryo provide preferential routes leading crest cells to the gut and to the suprarenal gland, respectively. There, the phenotypic expression of the crest cell population is regulated by environmental factors which elicit either cholinergic or adrenergic cell differentiation, irrespective of their fate in normal development.

It can be concluded that the potential for giving rise to cholinergic parasympathetic cells, adrenergic sympathetic ganglia, and adrenomedullary paraganglia is not locally restricted but is present in the entire crest.

The pluripotentiality of neural crest was also shown by Noden[13] by heterotopic transplantation of crest from the cranial level. The forebrain crest normally never gives rise to neurons, but when grafted at the mid/hindbrain region, crest cells emigrated from their new position and formed normal ciliary and trigeminal ganglia. These ganglia were, however, absent when the reverse transplantation (i.e., replacement of diencephalic by mesen/metencephalic crest) was performed.

Thus, the different regions of the neural crest are potentially capable of giving rise to both cholinergic and adrenergic cells, even if only one or the other type is produced under normal developmental conditions. Although the results of the transplantation experiments cannot provide rigorous proof that individual neural crest cells are bipotent in this respect, they show that the ultimate expression of the neuronal phenotype is highly dependent on extrinsic factors.

The next series of experiments shows that the role of the local environment on the differentiation of peripheral ganglion cells is not restricted to an early phase of development. Tissue environment obviously plays a role during a long developmental period and seems to insure the maintenance of the differentiated state as shown by Johnson et al.[14] on the superior cervical ganglion of the adult rat.

In these experiments,[15-17] instead of exchanging neural primordia before the onset of migration and differentiation of the neural crest cells, we back-transplanted into 2-day chick host crest derivatives which had achieved their normal migration and were in the process of differentiation. The tissues used were the ganglion of Remak, the ciliary ganglion, and the sympathetic ganglia or spinal ganglion, taken from 4- to 6-day quail embryos, i.e., at a stage when the ganglion cells already express their typical phenotype (for instance presence of CA or ACh). The graft was inserted close to the host's own neural crest, between the neural tube and somites, in the trunk region. Thus, differentiating crest-derived ganglia were subjected to the environment of the trunk of a younger embryo at the stage of neural crest cell migration and differentiation. The evolution of these supernumerary grafts was observed at various times after transplantation.[15, 17]

During the first hours after implantation, cells from the periphery of the grafted ganglion detached from the implant and started to migrate along the side of the host's neural tube. After 24 to 48 hrs, the cells of the graft were completely dispersed among the host's truncal tissues. If examined 4 to 6 days after grafting, the chimeric embryos showed that grafted quail cells were located exclusively in the normal sites of arrest of neural crest cells (ganglion of the PNS, nerves as Schwann cells, adrenal glands and plexuses). Although at the time of transplantation they had already undergone migration, the ganglion cells behaved as premigratory undifferentiated neural crest cells. They migrated anew and recognized the sites where neural crest cells normally stop.

Not only did the grafted ganglion cells locate at the normal sites of peripheral ganglion formation, but they also differentiated and expressed the phenotype corresponding to their environment: quail ciliary ganglion cells originating from a cholinergic population or sensory ganglion cells, when found in sympathetic ganglia or adrenal medulla of the host, responded positively to the formol-induced fluorescence method,[18] which means that they contained CA and had become adrenergic.

Spinal and autonomic grafted ganglia behaved differently in the host. Whereas the sensory ganglion-derived cells colonized not only in the host's dorsal-root ganglia but also in the sympathetic chain and adrenal medulla, the autonomic ganglion cells were located in the autonomic structures of the host only. This result indicates that sensory and autonomic ganglion cells have different affinities for the neural crest cell-arrest sites.

In a recent experiment,[19] the back-transplantation technique described above enabled us to demonstrate neuronal developmental potentials in the population of satellite cells of the nodose ganglion. In the nodose ganglion, the neurons are derived from the 3rd epibranchial placode, whereas the satellite cells are produced by the neural crest.[20] Because of this mixed origin, it is possible to construct chimeric nodose ganglia of the vagus nerve in which the satellite cells are labeled by the quail nucleolus, while the neurons are of the chick type by grafting a quail rhombencephalic primordium into a chick embryo early in development. Pieces of these ganglia were subsequently grafted in 2-day chick embryos into a slit located between the neural tube and the somites (for the technique see reference 15). The evolution of the graft (taken from the donor at 5.5–9 days of incubation) was similar to that observed for the other types of peripheral ganglia. The interesting point is that the quail cells were found not only in the host DRG (a few only) and in the ventral rachidian roots and branchial nerves such as Schwann cells, but also in the sympathetic ganglia, the adrenal medulla, and the enteric ganglia. Some of the quail cells located in the adrenergic host structures contained CA, and most cells located in the enteric plexuses differentiated into neurons. This shows that rhombencephalic neural crest cells, which do not normally contribute to the neuronal cell population in the nodose ganglion, can express such a phenotype if they are back-transplanted into the dorsal structures of a younger host. One can hypothesize that, in contact with the neurons of placodal origin, neuronal potentials are repressed in these cells, which then either remain undifferentiated or become engaged along the glial cell line differentiation pathway. As a result of the cell dissociation that takes place after the graft of the ganglion in the chick host, the quail cells become isolated and migrate. Some stop in the sympathetic ganglia and in the adrenal medulla where they are subjected to the environmental cues which lead them to express the adrenergic phenotype; others pursue their migration and reach the gut wall where they differentiate into enteric ganglia.

CONCLUDING REMARKS

The experiments described in this paper show that, although a regionalization can be recognized in the neural crest with respect to its fate as parasympathetic or sympathetic ganglia, considerable homogeneity actually exists in the developmental capabilities of the crest cell population along the neural axis. The final localization of the precursors of the PNS depends on the pathways they follow after leaving the neural primordium. The sites of arrest for the precursors of the ANS, in the gut or in the adrenal medulla for example, are recognized equally well by crest cells arising from any region of the neural axis. Plexuses of Meissner and Auerbach of the gut can be formed by cells of the cephalic or truncal level of the crest just as they can by their normal vagal precursors.

The cell recognition mechanisms that probably account for the architectural arrangement of the ANS in the truncal area are "memorized" in the crest-derived cells for a fairly long period during development. Quail ciliary or sympathetic ganglion cells grafted into a 2-day chick become distributed throughout the autonomic structures of the host and practically never form supernumerary ganglia located in abnormal sites. This is true even when the grafted ganglia are taken from embryos as late as 13 days of development (ciliary ganglion; see reference 21).

The experiments involving back-transplantations of peripheral ganglia of quail into 2-day chick embryos strongly suggest that differences in cell membrane properties exist between autonomic and sensory ganglion cells as early as 4.5 days of incubation. This conclusion is based on the respective localization of autonomic and DRG cells implanted into a younger host. The results observed suggest that restrictions of developmental capacities take place early in autonomic cells whereas a broader range of differentiation can still be displayed by DRG under the same conditions.

The experiment with the chimeric nodose ganglion revealed that inducible neuronal traits are present in the non-neuronal population of this ganglion as late as 9 days of incubation. Because of its mixed crest-placodal origin, the nodose ganglion is not strictly equivalent to the PNS ganglia of the trunk (which are entirely derived from the neural crest). This experiment raises a question concerning the interpretation of the results observed when the various other kinds of ganglia are back-transplanted into a younger host. In particular, when a developing cholinergic ganglion gives rise to adrenergic cells, the problem of whether the latter are derived from cholinergic neurons in which CA synthesis is induced or from a population of still undifferentiated cells is posed.

Acknowledgements

This work was supported by the Centre National de la Recherche Scientifique and grants from the Délégation Générale à la Recherche Scientifique et Technique and from the National Institute of Health (n° R01 DE0 4257 03 CBY).

REFERENCES

1. Tarkowski, A. K.: Mouse chimaeras developed from fused eggs. *Nature* (Lond.), **190**: 857–860, 1961.
2. Mintz, B.: Experimental study of the developing mammalian egg: Removal of the zona pellucida. *Science,* **138**: 594–595, 1962.
3. McLaren, A.: Mammalian Chimaeras. Cambridge University Press, Cambridge, 1976.
4. Le Douarin, N.: Particularités du noyau interphasique chez la caille japo-

naise (*Coturnix coturnix japonica*). Utilisation de ces particularités comme "marquage biologique" dans les recherches sur les interactions tissulaires et les migrations cellulaires au cours de l'ontogenèse. *Bull. Biol. Fr. Belg.*, **103**: 435–452, 1969.

5. Le Douarin, N.: A Feulgen-positive nucleolus. *Exp. Cell Res.*, **77**: 459–469, 1973.

6. Le Douarin, N. M.: Cell migration in early vertebrate development studied in interspecific chimeras. In: Embryogenesis in Mammals: Ciba Foundation Symposium 40, pp. 71–101. Elsevier/Excerpta Medica, North-Holland, Amsterdam, 1976.

7. Le Douarin, N. M.: Migration and differentiation of neural crest cells. In: Neural Development in Model System (ed. R. K. Hunt): Current Topics in Developmental Biology, Vol. 16, pp. 31–85. Academic Press, Inc., New York, 1980.

8. Le Douarin, N. M.: Plasticity in the development of the peripheral nervous system. In: Development of the Autonomic Nervous System: Ciba Foundation Symposium 83, pp. 19–50. Pitman Medical, London. 1981.

9. Noden, D. M.: Interactions directing the migration and cytodifferentiation of avian neural crest cells. In: The Specificity of Embryological Interactions (ed. D. Garrod), pp. 3–47. Chapman and Hall, London, 1978.

10. Le Douarin, N. and Teillet, M. A.: The migration of neural crest cells to the wall of the digestive tract in avian embryo. *J. Embryol. exp. Morphol.*, **30**: 31–48, 1973.

11. Le Douarin, N. and Teillet, M. A.: Experimental analysis of the migration and differentiation of neuroblasts of the autonomic nervous system and of neurectodermal mesenchymal derivatives using a biological cell marking technique. *Dev. Biol.*, **41**: 162–184, 1974.

12. Le Douarin, N. M., Renaud, D., Teillet, M. A. and Le Douarin, G. H.: Cholinergic differentiation of presumptive adrenergic neuroblasts in interspecific chimaeras after heterotopic transplantations. *Proc. Nat. Acad. Sci. U.S.A.*, **72**: 728–732, 1975.

13. Noden, D. M.: The control of avian cephalic neural crest cytodifferentiation, II: Neural tissues. *Dev. Biol.* **67**: 313–329, 1978.

14. Johnson, M. I., Iacovitti, L., Higgins, D., Bunge, R. P. and Burton, H.: Observations on the growth and development of sympathetic neurons in tissue culture. In: Development of the Autonomic Nervous System: Ciba Foundation Symposium 83, pp. 108–122. Pitman Medical, London, 1981.

15. Le Douarin, N. M., Teillet, M. A., Ziller, C. and Smith, J.: Adrenergic differentiation of cells of the cholinergic ciliary and Remak ganglia in avian embryo after *in vivo* transplantation. *Proc. Nat. Acad. Sci. U.S.A.*, **75**: 2030–2034, 1978.

16. Le Douarin, N. M., Le Lièvre, C. S., Schweizer, G. and Ziller, C.: An analysis of cell line segregation in the neural crest. In: Cell Lineage, Stem

Cells and Cell Determination (ed. N. Le Douarin), pp. 353–365. Elsevier/ North-Holland, Amsterdam, 1979.

17. Le Lièvre, C. S., Schweizer, G. G., Ziller, C. M. and Le Douarin, N. M.: Restrictions of developmental capabilities in neural crest cell derivatives as tested by *in vivo* transplantation experiments. *Dev. Biol.*, **77**: 362–378, 1980.

18. Falck, B.: Observations on the possibilities of the cellular localization of monoamines by a fluorescence method. *Acta Physiol. Scand.* **56**, *suppl.*, **197**: 1–25, 1962.

19. Ayer-Le Lievre, C. S. and Le Douarin, N. M.: The early development of cranial sensory ganglia and the potentialities of their component cells studied in quail-chick chimaeras. *Dev. Biol.* (in press)

20. Narayanan, C. H. and Narayanan, Y.: Neural crest and placodal contributions in the development of the glossopharyngeal-vagal complex in the chick. *Anat. Rec.*, **196**: 71–82, 1980.

21. Ziller, C., Smith, J., Fauquet, M., and Le Douarin, N. M.: Environmentally directed nerve cell differentiation: *in vivo* and *in vitro* studies. In: Development and Chemical Specificity of Neurons (ed. M. Cuénod, G. W. Kreutzberg and F. E. Bloom): Progress in Brain Research, Vol. 51, pp. 59–74. Elsevier/North-Holland, Amsterdam, 1979.

III. MAMMALIAN MUTANTS

Examples of Neuromuscular Mutations in the Mouse: Opportunities for an Understanding of Pathological Processes*

Hans Meier

We have been studying neuromuscular mouse mutants for the last fourteen years for two basic reasons: to learn more about the mode of earliest gene action and to compare and evaluate them as models of human neuromuscular diseases. The latter included clinical, descriptive pathological, and biochemical studies as well as abnormal structure-function relationships. This paper will also briefly describe our present approaches to comprehend gene action and gene expression.

With regard to neuromuscular mutants the general method which we employed was that of retrograde analysis to determine the primary cause(s) of the neuro- or myopathy. Also, we have been able to produce litters containing normal and dystrophic animals distinguishable by an eye color marker which allowed us to study the pathogenesis of the *dy*-muscular dystrophy. Clearly, studying the various developmental single-gene effects of neuromuscular mutations is only a limited approach to analyzing a highly complex process. What directs the differentiating cells to migrate into proper positions and what specific interactions between surface molecules stably organize cells into an organ are questions still unanswered. On the other hand, mutant genes with multiple, pleiotropic effects or independent loci may be useful for revealing relationships between apparently diverse developmental processes. For example, there are at least six independent loci which affect the mouse cerebellum:[1] weaver and staggerer result in a loss of granular cell neurons; reeler causes a disarray of alignment and an impairment of the synoptic relationships among neurons of the cerebellar cortex; leaner reveals focal degeneration of granular and Purkinje cells in the anterior vermis; swaying causes an embryonic malformation of the anterior vermis; and nervous causes a selective loss of neurons of the cerebellar cortex.

Another approach to the analysis of the CNS is the use of chimeras (mix-

The Jackson Laboratory, Bar Harbor, Maine, U.S.A.

* Dr. Meier could not attand the meeting due to illness.

ing separate cells of an early normal embryo with cells of a mutant followed by reimplantation in a host uterus).

Finally, to understand gene expression Neal G. Copeland and Nancy A. Copeland in my laboratory use recombinant DNA methodology to recognize active and inactive genes, gene amplification, and deletion that occur in some differentiated and transformed cells. Briefly, our long term objectives are to understand the mechanisms controlling normal eukaryotic gene expression and how they are affected during enoplastic transformation. A useful approach for studying these questions involves assaying the biological activity of retroviral and cellular DNAs by DNA transfection. Recently, we have used assays of the biological activity of Rous sarcoma virus (RSV) DNA, by transfection of NIH 3T3 mouse cells, to study mechanisms of viral transformation and gene regulation.

The objectives of this study are: to characterize, at the molecular level, the mechanism of transfection of NIH 3T3 cells by RSV DNA; to determine the mechanism(s) of generation of defective proviral DNAs found in several NIH 3T3 cell lines transformed by transfection of unintegrated RSV DNA; to understand how cellular DNA sequences can alter and/or control viral gene expression; and to study the effect of proviral DNA sequences on transcription of flanking cellular genes.

Specifically, our approach to answering these questions included cloning into the Hind III site of bacteriophage lambda (Charon 9) the proviral DNAs and flanking cellular DNA sequences of several NIH 3T3 cell lines transformed by transfection of unintegrated, integrated, and subgenomic fragments of RSV DNA. Recombinant DNA clones and RSV DNA transformed NIH 3T3 cells were then characterized by restriction enzyme digestion, Southern blotting, and DNA transfection. This enabled us to analyze the structure of the defective proviral DNAs found in NIH 3T3 cells transformed by intact and subgenomic fragments of RSV DNA and to determine the origin, arrangement, and structure of cellular DNA sequences flanking the integrated provirus. To study the effects of proviral DNA sequences on transcription of flanking cellular genes, hybridization probes specific for flanking cellular DNA sequences were prepared from recombinant DNA clones and were used to analyze RNAs synthesized in untransformed and RSV DNA transformed NIH 3T3 cells. Similar probes and approaches can be applied to problems in neuromuscular diseases and mammalian development, but this is yet in the future.

Studies have been reported concerning the development and availability of teratocarcinoma stem cells. These reports have made possible new biochemical, immunological, and genetic studies of early embryonic differentiation. In addition, the production of teratocarcinoma-embryo chimeras may ultimately lead to the creation of strains of mice with novel genotypes, some of which may serve as models of human disease.

We believe that mouse neurological disease models, even though we found them to be generally more complex than human hereditary neurological diseases, have provided important insights into nervous system abnormalities. For example, the mutations teetering[2] and ducky[3] have contributed to a better understanding of the functional role of the pyramidal tract, despite the facts that the pyramidal tract is a phylogenetically new acquisition inherent only to man, and that the major impulses in animals are transmitted through other tracts and from segment to segment of the spinal cord. Both the teetering and ducky mutants are characterized by corticospinal tract dysgenesis, although teetering mice suffer from progressive paralysis, whereas paralytic signs are absent in ducky. I consider these findings important positive results.

However, it has been extremely difficult to pinpoint the earliest developmental (etiological) changes which lead to the various neurological diseases even when genetic markers are available to identify mutants prior to the occurrence of symptoms, e.g. in ducky and muscular dystrophy. In addition, mutants are often small and runted compared to their normal littermates and have a considerably reduced lifespan.

Thus, as we have learned more about the characteristics of the mutant-bearing mice studied, we have found valuable parallels and differences between human and mouse diseases, making the mouse disorders more useful as models for analysis and experimentation.

Ducky (gene symbol, *du*) is a locomotor ataxia caused by spinomedullocerebellar dysgenesis, and demyelination of selected fiber systems, including auditory pathways. There is a major deficiency of cerebrosides. Teetering (*tn*) is a condition caused by deficiency and underdevelopment of selective central nervous system (CNS) regions. Spastic (*spa*) is characterized by intervertebral arthropathy and enchondrosis resulting in part from disc herniations and ossifications, as well as leptomeningeal cyst formation.[4] Leaner (*ln*) is a so-called cerebellar mutant because of severe abnormalities involving mainly the cerebellum. Dr. S. Tsuji found, when he worked in my laboratory, that the leaner gene (*ln*) is allelic with tottering (*tg*), located on Chromosome 8 of the mouse, and is closely linked to the esterase-1 (*Es-1*) locus.[5,6] Tottering is characterized by epileptiform seizures and the compound heterozygotes (*tg/ln*) have symptoms of both.[7] Myelin synthesis deficiency (*msd*), which has now been determined to be an allele of jimpy (*jp*) and designated *jp^{msd}*, causes tremors, repeated seizures, and death of affected males at about three weeks of age.[8-10] The CNS is markedly deficient in myelin. Thus, severe disturbances in myelin metabolism occur in at least three neurological mutants, quaking (*qk*), jimpy (*jp*) and myelin synthesis deficiency (*jp^{msd}*). Major abnormalities in all three mutants involve cerebrosides and sulfatides. The basic defects are presumed to reside in the oligodendroglia of the CNS, since the peripheral nerves appear to be properly myelinated. Quaking and jimpy mutants also have an altered polyamine synthesis and accumulation, particularly of the hindbrain and spinal cord. *The finding of genes located on different chromosomes that cause similar disorders of the CNS is of considerable importance since several genes may influence the same gene pathways in CNS metabolism and development.* A similar situation pertains to dystrophia muscularis (*dy* and *dy^{2J}*) located on Chromosome 10 and myodystrophy (*myd*), mapped to Chromosome 8 of the mouse genome, that cause similar disorders of the skeletal musculature.[11] Recently, Ms. P. W. Lane of the Jackson Laboratory has identified a fourth muscular dystrophy which is genetically related to neither *dy* nor *myd*, although I have observed many pathological features in common between them. Our studies indicated that comparison of the three hereditary myopathies caused by the first three mutant genes may help explain the fact that "muscular dystrophy" in man defines a group of disorders having both similar and individual characteristics. We have also found that multiple or pleiotropic gene effects as well as interactions between genes may occur not only in mice but probably also in man. Specifically, we found interaction of *dy^{2J}* with dominant spotting (*W*) and steel (*Sl*). These two genes have been found to cause anemia, sterility, and dilution of pigment and causing morphological and possibly functional mitochondrial changes,[12] skeletal muscle changes include severe sarcoplasmic-vescicular enlargements and dissolution, mitochondrial swelling with loss of cristae, and a compacting and fragmentation of Z-band myofibrils in some fibers, whereas other fibers remain entirely normal. The pathogenesis of the lesions caused by both *Sl* and *W* are yet unknown. As a matter of fact, it is not known whether the primary gene defect in each of the myopathic mutations is muscular or neural. In *myd* mutants we found extremely high levels of calcium in their diaphragms, while certain other muscles have moderately but inconsistently elevated calcium. These calcium deposits may ultimately have been derived from the skeleton. The calcium contents

were not consistently altered from those of non-mutant littermates in both
dy and *dy²ᴶ* mutants.[13]

In a recessive mutation called muscle deficient (*mdf*), we observed
progressive reduction of muscle mass in the hind limbs.[13] This muscular
atrophy is probably secondary to a peripheral neuropathy. The condition
is neither sex-limited nor sex-linked, and the mutation is non-allelic with
eleven other neurological and muscular mutations. Linkage of *mdf* remains
elusive, segregating independently of 25 markers on 12 different mouse
chromosomes.[14]

Finally, shambling (*shm*) is an autosomal recessive mutation of the
mouse that appeared in the fifth generation of a genetically heterogeneous
population whose male ancestors had been exposed to gonadal radiation
in each of the preceding 4 generations. This randombred population was
derived from a generation of double cross hybrids of inbred strains
C57BL/6J, BALB/cJ, DBA/2J, and C3HeB/FeJ. The mutant mice, two
females in a litter of 6, were slightly smaller than their sibs at weaning,
dragged their hind feet when walking, swayed from side to side, and
trembled severely when disturbed. It was found that the shambling
neuropathy is distinct from numerous other hereditary neurological conditions reported in mice. In this disease, alcian blue and PAS-positive
deposits occur throughout the dorsal roots of the lumbar spine. It is a
sensory disorder of probable metabolic origin in which the most severe
involvement is in the lumbar dorsal roots.

Many problems still remain unsolved. The most basic question awaiting
an answer concerns the mechanism of faulty gene action. In order to
determine the primary gene effects, pathological studies have been initiated, retrogressing toward earlier stages in the life of mutant mice.[15]
Obviously, knowledge of the types and sites of established lesions is a
prerequisite for pathogenic studies.

Based on my experience at the Jackson Laboratory, we may expect
about 25 presumptive neurological mutants per one million mice examined. A vast number of them are now available for study and experimentation. I believe that we have made significant progress towards
characterizing many of them and utilizing them to obtain basic information about normal and abnormal development, differentiation, and
the structure and function of component nervous and muscular systems.
However, much still remains to be accomplished.

In summary, we have attempted to elucidate the pathogenesis of selective hereditary neuromuscular diseases in mice for the last 14 years. The mutations to be studied were selected on the basis of their potential value with respect to their genetical and neuropathological features. We hoped to compare them with and evaluate their usefulness as models of human hereditary neurological and neuromuscular diseases. Even though they have been found to be generally more complex than human hereditary neurological diseases, they have provided important insights into nervous system abnormalities. We found it extremely difficult to pinpoint the earliest developmental (etiological) changes which lead to the various neurological diseases even when genetic markers are available to identify mutants prior to the occurrence of symptoms.

Our second major aim in studying neuromuscular mutants was to understand the mechanisms controlling normal and abnormal eukaryotic gene expression and how they affect CNS development and differentiation. Although this focus is still of major interest our newest approach in answering questions regarding how cellular DNA sequences can alter or control gene expression has now been drastically changed. The questions with regard to gene action relate to the site or cell type and the synaptic effects. Much is still to be gained in this area. Certainly we do not as yet understand the control of any single eukaryotic gene with molecular detail nor know its exact mode of action.

REFERENCES

1. Sidman, R. L. and Green, M. C.: "Nervous", a new mutant mouse with cerebellar disease. *Colloques Internationaux au Centre National de la Recherche Scientifique* No. 924, pp. 69–79.
2. Meier, H.: The neuropathy of teetering, a neurological mutation in the mouse. *Arch. Neurol.*, **16**: 59–66, 1967.
3. Meier, H.: The neuropathology of ducky: A pathological and preliminary histochemical study. *Acta Neuropathol.*, **11**: 15–28, 1968.
4. Meier, H. and Chai, C. K.: Spastic, a hereditary neurological mutation in the mouse, characterized by vertebral arthropathy and leptomeningeal cyst formation. *Exp. Med. Surg.*, **28**: 24–28, 1971.
5. Tsuji, S. and Meier, H.: Evidence for allelism of leaner and tottering in the mouse. *Genet. Res.*, **17**: 83–88, 1971.
6. Tsuji, S. and Meier, H.: Linkage of serum esterase and tottering in the mouse. *J. Hered.*, **60**: 221–222, 1969.
7. Meier, H. and MacPike, A. D.: Three syndromes produced by two mutant

genes in the mouse: Clinical, pathological and ultrastructural bases of tottering, leaner, and heterozygote mice. *J. Hered.*, **62**: 297–302, 1971.

8. Meier, H. and MacPike, A. D.: A neurological mutation (*msd*) of the mouse causing a defiiciency of myelin synthesis. *Exp. Brain Res.*, **10**· 512–525, 1970.

9. Brenkart, A., Arora, R. C., Radin, N. S., Meier, H. and MacPike, A. D.: Cerebrosides synthesis and hydrolysis in a neurological mutant mouse (*msd*). *Brain Res.*, **36**: 195–202, 1972.

10. Day, E. D., Mickey, D. D., Rigsbee, L. C. and Meier, H.: Zonal centrifugation and flotation-fractionation of *msd* mutant mouse brain. *J. Neurobiol.*, **3**: 325–335, 1972.

11. Meier, H. and MacPike, A. D.: Myopathies caused by three mutations of the mouse. *J. Hered.*, **68**: 383–486, 1977.

12. Meier, H. and MacPike, A. D.: Pleiotropic gene effects on muscle ultrastructure of normal and dystrophic mice. *Exp. Neurology*, **40**: 258–262, 1973.

13. Nutting, D. F., MacPike, A. D. and Meier, H.: The calcium content of various tissues from myodystrophic and dystrophic mice. *J. Hered.*, **71**: 15–18, 1980.

14. Womack, J. E., MacPike, A. D. and Meier, H.: Muscle deficient (*mdf*), a new mutation in the mouse. *J. Hered.*, **71**: 68, 1980.

15. Meier, H.: Pathological findings in shambling, a hereditary neuropathy of mice. *J. Neuropathol. Exp. Neurol.*, **26**: 620–633, 1967.

A Study of Myelin Deficiency in the Central Nervous System of Mice with Genetic Muscular Dystrophy

Shigekatsu Tsuji

INTRODUCTION

Muscular dystrophy in the mouse was originally considered to be a genetically determined primary disorder of skeletal muscle exhibiting a progressive necrotizing myopathy, and hence to be a good model of a certain form of human muscular dystrophy.[1] Subsequent studies, however, have revealed morphological and physiological abnormalities in the peripheral nervous system of these animals. Morphological studies have shown a reduction in the number of peripheral nerve fibers,[2,3] structural alterations in motor nerve terminals,[4] and deficiency of the myelin sheath in the spinal and cranial roots.[5,6] The studies of physiological consequences of neural abnormalities in dystrophic mice have also demonstrated the functional denervation in large numbers of muscle fibers of tibialis anterior and gastrocnemius muscles.[7] In contrast, there are relatively few studies of the central nervous system in dystrophic mice compared to the number of studies on the skeletal muscle and peripheral nerves.

The present study was undertaken to biochemically examine the brain and spinal cord of dystrophic mice at an early stage of the disease. In addition, we hoped to determine if membrane abnormalities exist in the central nervous system, and to clarify the relationship between the neural abnormality and the development of muscular dystrophy.

MATERIALS AND METHODS

Animals

C57BL/6J-*dy* strain mice originally obtained from The Jackson Laboratory (Bar Harbor, Maine, U.S.A) in 1970 were bred in our laboratory. Experiments were carried out simultaneously with homozygous affected

Department of Physiology, Wakayama Medical College, Wakayama, Japan

(dy/dy) and unaffected littermates ($+/?$) at various ages. Dystrophic mice were identified by observation of periodic dragging of the rear feet and clasping of the hindlimbs when the animal was lifted by the tail beginning 12–14 days after birth. The brain and spinal cord were excised for subcellular fractionation, lipid analysis, and enzyme assays.

Subcellular fractionation and total lipid determination

The brain or spinal cord was homogenized in ten volumes of 0.32 M sucrose and centrifuged for ten minutes at 900 × g in a refrigerated centrifuge. The resulting precipitate (Pl) was washed four times with 0.32 M sucrose solution. The supernatant (Sl) and the washing solutions were combined and centrifuged for 30 minutes at 11,500 × g. After washing once with 0.32 M sucrose, the resulting precipitate (P2) was subjected to osmotic shock in redistilled water and centrifuged for 30 minutes at 20,000 × g. Then the resulting precipitate (Ml) was resuspended in 0.32 M sucrose and used for the subfractionation. Subfractionation with a discontinuous sucrose gradient was performed using transparent tubes for the swing rotor according to the procedure of Lapetina et al.[8]

Extraction of lipid components of the brain and spinal cord was performed according to the Folch method.[9] After washing with 0.37% KCl solution and with the solution of pure solvent upper phase, the Folch extracts were used to determine the total lipid content. The total phospholipid content was determined by measuring the amount of inorganic phosphorous released by digestion.[10] Total cholesterol content was analysed by the Zak method.[11] Determination of protein content in the proteolipids was performed by the Lowry method[12] on a lipid extract after hydrolysis by NaOH. Total cerebroside content was measured using hexose portions with an alternative micromethod using the orcinol reaction.[13] Sulfatides were determined separately by the Azure A colorimetric assay method.[14] Estimation of sphingosine was performed by the procedure of Lauter and Trams.[15]

Enzyme assays

The tissues were homogenized in ten volumes of ice cold 0.32 M sucrose in a teflon homogenizer. The purifed myelin fraction was isolated with a conventional subcellular fractionating procedure, essentially according to Norton,[16] and used as an enzyme source after pretreating with 0.3 mg/ml Triton x-100 solution.

For the determination of 2′,3′-cyclic nucleotide 3′-phosphohydrolase (CNP) activity, two methods were employed: the method of Olafson et *al.*[17] with 2′,3′-cyclic AMP as substrate, and the method of Sogin[18] with 2′,3′-cyclic NADP as substrate.

To assay the cholesterol ester hydrolase (CEH) activity in the purified myelin fraction, 250 μg cholesterol oleate in benzene and 2 μmoles of sodium taurocholate in methanol were dried together under nitrogen and suspended in 1.3 ml of 0.1 M sodium phosphate buffer (pH 7.2) containing 3 mg of bovine albumin by sonication for 30 sec. at 4°C. The substrate mixture was then added to 0.2 ml of the purifed myelin preparation and incubated for 20 minutes at 37°C with shaking. After 20 minutes of incubation, the reaction was stopped by the addition of 3.5 ml of iso-propanol in order to determine the free cholesterol content by the method of Richmond.[19]

Administration of protease inhibitors

Leupeptin and bestatin produced by actinomycetes were kindly supplied by Dr. H. Umezawa (Institute of Microbial Chemistry, Tokyo, Japan). All of the dystrophic mice used in this experiment were divided into two groups. One group was injected subcutaneously with 0.2 ml of protease inhibitor solution which contained 0.2 mg of leupeptin or bestatin, twice a day during various periods, and the control group was injected with saline instead of the protease inhibitor solution.

RESULTS

Isolation of subcellular fractions

A typical result of the ultracentrifugal pattern of crude mitochondrial fraction (Ml) isolated by a discontinuous sucrose gradient is illustrated in Fig. 1. There was a visible reduction in the spinal cords of dystrophic mice at 4 weeks of age when compared with the spinal cords of normal littermates.

Quantitative analysis of lipid extraction from the brain and spinal cord

The quantitative analysis of the principal lipid classes in the brain and spinal cord was then performed to find out if the myelin insufficiency of dystrophic mice involves an inborn error in the metabolism of one of its constituents. There was a marked decrease in the amount of total lipids extracted from both the brain (362.9 ± 21.6, $n = 12$) and spinal cord (132.1 ± 8.3, $n = 12$) of dystrophic mice compared to that of normal control mice (396.1 ± 26.0 and 146.2 ± 8.1, respectively, $n = 10$).

The results of the quantitative analysis of lipid composition in the spinal

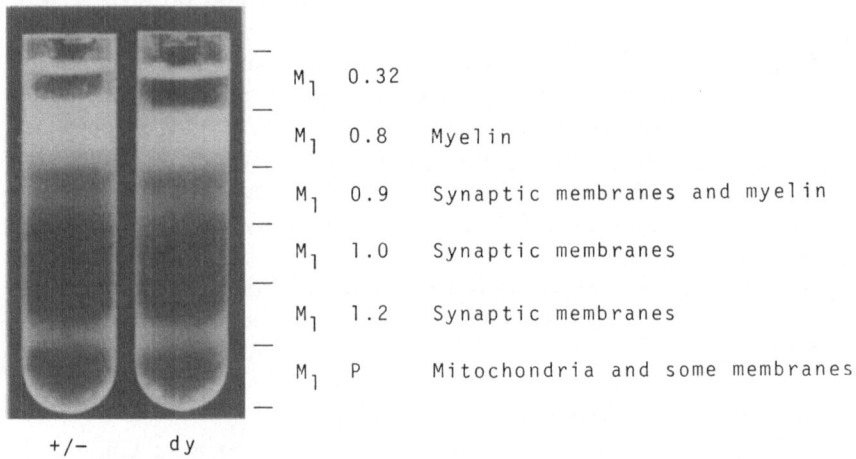

FIG. 1. Ultracentrifugal pattern on a discontinuous sucrose gradient of M1 fraction isolated from spinal cords of dystrophic mice and normal controls.

TABLE 1.   Lipid Composition of Spinal Cords in Dystrophic and Normal Mice

	Dystrophic	Normal	D/N × 100
Cholesterol (mg)	21.46 ± 7.08	16.18 ± 1.21	132.6
Proteolipids (mg protein)	6.63 ± 3.12	7.61 ± 3.94	87.1
Phospholipids (mg P)	3.16 ± 0.20	3.92 ± 0.27	80.6
Cerebrosides (mg)	15.42 ± 1.68	25.37 ± 6.88	60.8*
Sphingosine (arbitrary unit)	29.0 ± 0.1	34.3 ± 10.9	84.5
Sulfatides (μ moles)	3.13 ± 0.05	4.35 ± 0.28	71.9*

All units are based on g wet tissue weight.
*; $p < 0.05$

TABLE 2.   Brain Lipids Composition of Dystrophic and Normal Mice

	Dystrophic	Normal	D/N × 100
Cholesterol (mg)	22.02 ± 3.14	22.18 ± 4.84	99.3
Proteolipids (mg protein)	5.26 ± 1.43	5.03 ± 1.90	104.6
Phospholipids (mg P)	2.37 ± 0.85	2.28 ± 0.87	103.9
Cerebrosides (mg)	8.75 ± 2.32	8.79 ± 3.20	99.5
Sphingosine (arbitrary unit)	10.7 ± 4.4	12.4 ± 4.9	86.3
Sulfatides (μ moles)	3.64 ± 1.68	4.71 ± 1.50	77.3*

All units are based on g wet tissue weight.
*; $p < 0.05$

cord and brain are summarized in Tables 1 and 2. All of the lipid classes described here, except cholesterol, were reduced in the spinal cord of dystrophic mice (Table 1). Cerebrosides and sulfatides, which are major constituents of myelin, diminished significantly in dystrophic mice compared to normal mice, while the amount of other lipids showed no significant changes. It is noteworthy that the total cholesterol content in the spinal cords of dystrophic mice was greater than that of the controls.

A significant decrease of sulfatides concentration was also observed in the brain of dystrophic mice when compared with those of normal mice (Table 2).

Myelin CNP and CEH activities

In order to elucidate the disturbances of myelin metabolism in dystrophic mice, we investigated the developmental changes of 2′,3′-cyclic nucleotide 3′-phosphohydrolase (CNP) and cholesterol ester hydrolase (CEH) activities in the purified myelin fraction from the central nervous system. These two enzymes are known to be closely associated with the myelin structural protein.[20, 21]

There were significant reductions of total myelin CNP activity in the brain and spinal cord of dystrophic mice at 8 weeks of age (Table 3). However, the specific activities of the enzyme based on mg of protein were comparable in dystrophic mice and normal controls. The activities of myelin CEH in the brain and spinal cord of dystrophic mice and control mice at 8 weeks of age are summarized in Table 4. Significant reductions of both total and specific activities of myelin CEH were observed in the brain and spinal cord of dystrophic mice compared to those of normal mice.

TABLE 3.   Myelin 2′,3′-cyclic Nucleotide 3′-phosphohydrolase in Brains and Spinal Cords of Dystrophic Mice

	Dystrophic	Normal	D/N × 100	Pr.
Brain				
Total activity	11.52 ± 0.90	17.28 ± 1.63	66.7	$p < 0.01$
Specific activity	5.96 ± 1.10	6.47 ± 1.56	92.1	N.S.
Spinal cord				
Total activity	5.18 ± 0.72	7.60 ± 1.15	68.1	$p < 0.01$
Specific activity	4.14 ± 0.83	4.20 ± 1.11	98.6	N.S.

Unit; μ moles NADP/total protein or mg protein/min.
Averaged age of the mice is 51.3 (41–60) days.

TABLE 4. Myelin Cholesterol Ester Hydrolase in Brains and Spinal Cords of Dystrophic Mice

	Dystrophic	Normal	D/N × 100	Pr.
Brain				
Total activity	5.30 ± 0.26	9.52 ± 3.43	55.7	$p < 0.01$
Specific activity	2.09 ± 0.93	4.01 ± 1.01	52.1	$p < 0.01$
Spinal cord				
Total activity	4.94 ± 0.98	10.75 ± 2.57	45.9	$p < 0.01$
Specific activity	1.25 ± 0.44	2.26 ± 0.41	55.3	$p < 0.01$

Unit; mg cholesterol/total protein or g protein/min.
Averaged age of the mice is 57.8 (41–75) days.

The total activity of myelin CNP in the central nervous system of control mice was extremely low during the first 8 postnatal days, after which time there was a rapid increase lasting until days 28 (Fig. 2). A slight decline in the activity appeared to occur from this period until adulthood. In the spinal cord of dystrophic mice, the myelin CNP activity was extremely depressed, approximately 60% of the control activity at the maximum level. The response peaked at around 30 days after birth in both dystrophic and control mice. Even at 14 days of age, immediately after the disease appeared, the total activity in the spinal cord of dystrophic mice was only 70% of normal activity (Fig. 2b).

In the brains of dystrophic mice, the pattern of the increase in total myelin CNP activity was identical to that of the controls during the first 28 postnatal days. After this period, a sudden decrease in the activity occurred in the brains of dystrophic mice, dropping to approximately 60% of the control level at around 40 days of age (Fig. 2a). The reduced activity of total CNP in the central nervous system of dystrophic mice seems to be proportional to a reduced amount of total myelin membrane rather than to a specific defect in the synthesis of the enzyme protein.

A rapid increase in the activity of myelin CEH in the developing spinal cord of normal mice occurred during the period from 8–20 days, reaching an almost 20-fold increase over the newborn level. After this period, a gradual but steady decrease occurred until adulthood. The myelin CEH activity in dystrophic mice was depressed to approximately 60% of the control activity immediately after the disease appeared and reached only half of the normal activity at adulthood (Fig. 3b).

A rapid increase in the activity of myelin CEH occurred in the brain of control mice during days 8–30 (Fig. 3a). After this period, the activity remained constant or possibly declined slightly until adult level. In the brain

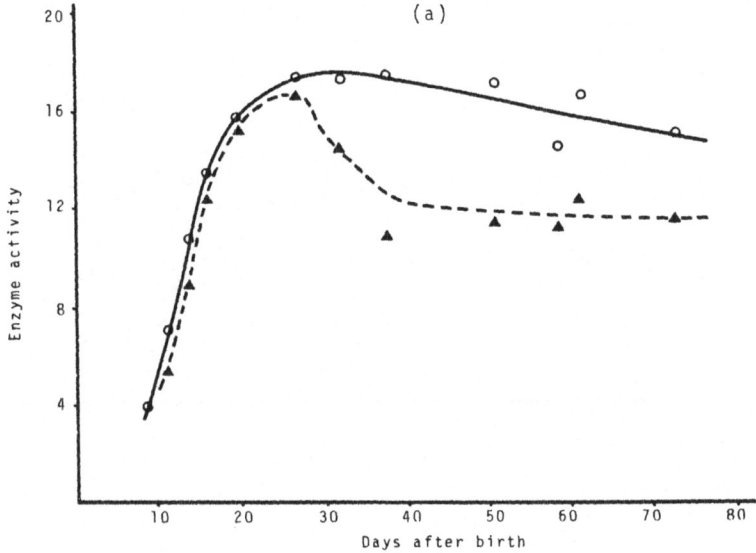

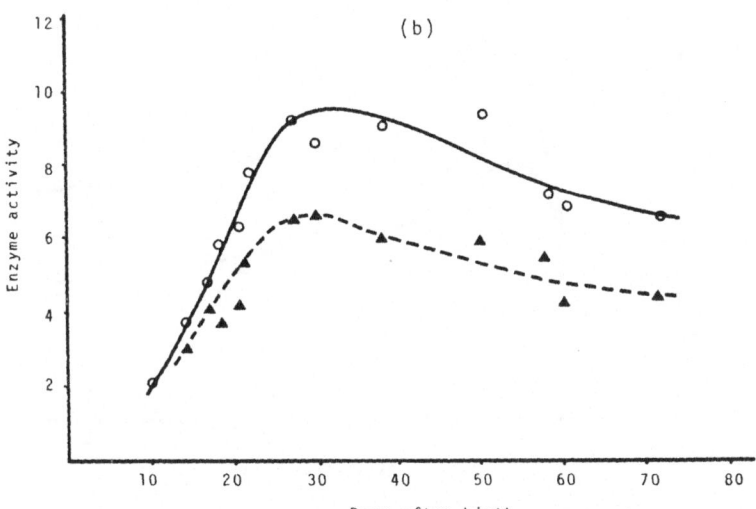

FIG. 2. Myelin CNP in the developing and maturing brains (a) and spinal cords (b) of dystrophic mice (▲- - -▲) and normal controls (O—O). Values are expressed as micromoles of NADP released per total myelin per minute.

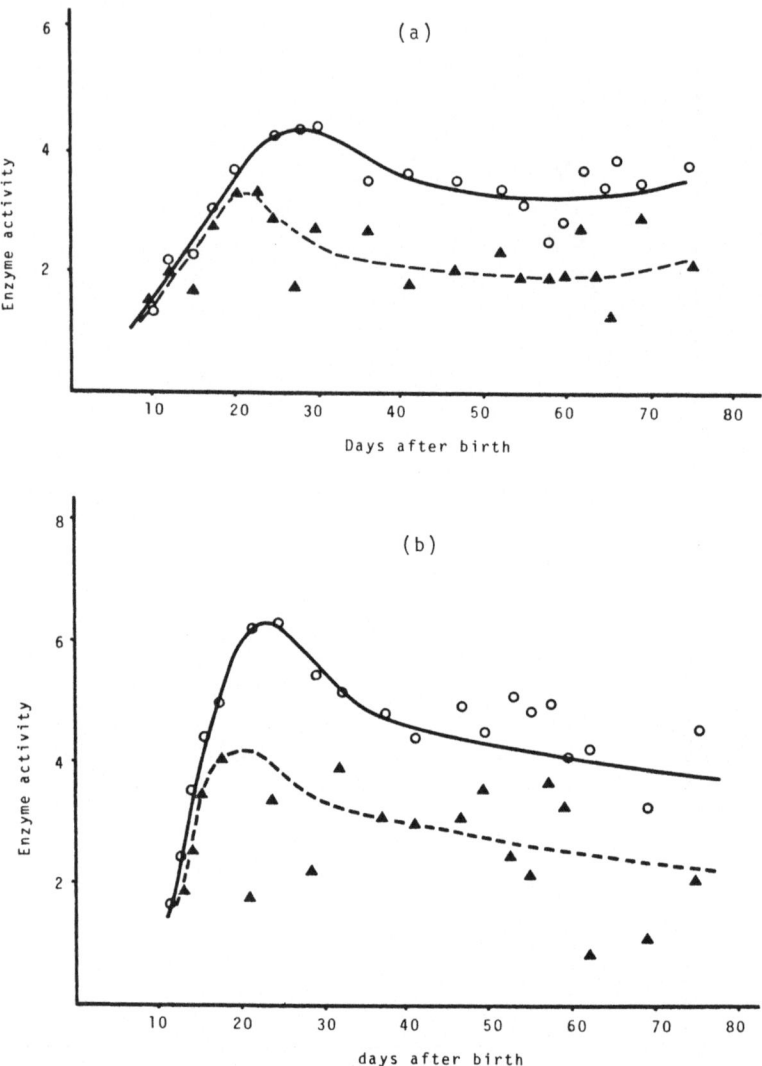

FIG. 3.   Myelin CEH in the developing and maturing brains (a) and spinal cords (b) of dystrophic mice (▲- - -▲) and normal controls (○—○). Values are expressed as mg of cholesterol released per g of protein per minute.

of dystrophic mice, the pattern of increase of myelin CEH was identical to that of normal littermates during the first 20 postnatal days, after which time a sudden decrease occurred lasting until 30 days dropping to approximately one-half of the control level. The findings concerning the correlation between the reduced activity in myelin CEH and the reduced synthesis of myelin suggests that this enzyme plays an important role in the development of hypomyelination in the central nervous system of dystrophic mice.

Effects of protease inhibitors on myelin associated enzymes

There have been several studies suggesting that the increased activities of some proteases in muscle and other tissues are causally related to muscular dystrophy in the mouse. In order to find the primary factor which leads to myelin abnormalities and its relation to the cause of muscle atrophy, we investigated the effects of administering protease inhibitors (leupeptin and bestatin) on the activity changes of myelin associated enzymes. Leupeptin is known to suppress the breakdown of a myelin structural protein[22] and bestatin is known to be an inhibitor of membrane bound aminopeptidases.[23]

The results are shown in Tables 5 and 6. There were striking recoveries in the activities of both myelin CNP and CEH in the spinal cord (Table 5) and brain (Table 6) of dystrophic mice treated with protease inhibitors. The influence of leupeptin on the activities of both enzymes in the spinal cord seemed to be more effective than that of bestatin (Table 5). In the bestatin-treated dystrophic mice, the recovery rate of both enzyme activities seemed to be proportional to the administration period (Table

TABLE 5. Effects of Protease Inhibitors on the Activities of Myelin-associated Enzymes in Spinal Cords of Dystrophic Mice

	Treatment	CNP	T/C × 100	CEH	T/C × 100
Control		4.14 ± 0.29		2.26 ± 0.75	
Leupeptin	2 weeks	10.37 ± 1.69	250.4*	9.43 ± 1.77	417.3*
Bestatin	2 weeks	8.40 ± 0.87	203.1*	2.35 ± 1.07	104.0***
Bestatin	4 weeks	8.15 ± 0.76	196.9*	4.29 ± 0.63	169.8**
Leupeptin + bestatin	4 weeks	7.61 ± 0.83	184.0*	2.32 ± 0.08	102.7***

All mice were sacrificed at 8 weeks of age and prepared for enzyme assay.
Units: CNP; μ moles NADP/mg protein/min, CEH; mg cholesterol/g protein/min.
*; $p < 0.001$, **; $p < 0.01$, ***; N.S.

TABLE 6. Effects of Protease Inhibitors on the Activities of Myelin-associated Enzymes in Brains of Dystrophic Mice

	Treatment	CNP	T/C × 100	CEH	T/C × 100
Control		5.93 ± 0.85		4.01 ± 0.44	
Leupeptin	2 weeks	12.86 ± 0.49	216.8*	7.87 ± 1.95	196.3**
Bestatin	2 weeks	14.25 ± 1.22	240.3*	4.57 ± 1.06	114.0***
Bestatin	4 weeks	15.68 ± 1.89	264.4*	14.25 ± 2.35	355.4*
Leupeptin + bestatin	4 weeks	14.17 ± 1.84	239.0*	7.25 ± 2.05	180.8***

All mice were sacrificed at 8 weeks of age and prepared for enzyme assay.
Units: CNP; μ moles NADP/mg protein/min, CEH; mg cholesterol/g protein/min.
*; $p < 0.001$, **; $p < 0.01$, ***; N.S.

6). The effects of leupeptin on these enzyme activities appeared almost immediately, while the effects of bestatin occurred more slowly.

Endurance test for the bestatin-treated mice

Continuous administration of bestatin to dystrophic mice at an early stage of the disease was performed to find the recovery effects of this protease inhibitor on the clinical manifestation of muscular dystrophy. During the first ten days of consecutive administration, all recipient mice showed normal growth and the disappearance of dragging of the hindlegs. After this period, however, two distinct groups developed among the bestatin-treated mice with regard to body weight. About one-half of the treated mice reached normal body weight, while the other half failed to gain much body weight. Table 7 compares the body weights of bestatin-treated and non-treated dystrophic mice at one month and two months

TABLE 7. Comparison of Body Weight of Bestatin-treated Dystrophic Mice with That of Untreated Mice

Sex	Treatment	1 month	2 month
Male	Untreated	8.5 ± 1.4 [6.5–10.0] (10)	10.0 ± 1.9 [6.5–13.5] (14)
	Treated	11.5 ± 2.7 [7.0–16.0] (20)**	16.5 ± 4.3 [11.5–27.5] (22)*
		9.8 ± 1.6 [7.0–11.5] (12)	13.1 ± 0.9 [11.5–14.5] (10)*
		14.1 ± 1.5 [12.5–16.0] (8)*	19.4 ± 3.8 [15.5–27.5] (12)*
Female	Untreated	6.8 ± 1.0 [6.0–8.0] (10)	10.9 ± 0.8 [10.0–12.0] (11)
	Treated	11.0 ± 2.3 [9.0–18.0] (14)*	13.5 ± 2.5 [8.5–20.5] (16)**
		9.7 ± 0.6 [9.0–10.5] (8)*	11.5 ± 1.6 [8.5–12.5] (10)
		12.8 ± 2.6 [11.0–18.0] (6)*	16.3 ± 2.5 [13.5–20.5] (6)**

Brackets show the range of each data. Parentheses show the numbers of mice used.
Unit: gm body weight.
*; $p < 0.001$, **; $p < 0.01$

of age. Even at one month of age, the average body weight of bestatin-treated mice was significantly higher than that of the non-treated mice.

Endurance tests for the skeletal muscle of bestatin-treated and weight-regained dystrophic mice and control dystrophic mice at two months of age were then performed to find out if the regaining of body weight was correlated with the recovery of muscle strength. The durability of the muscle was measured using a cylindrical wire cage on a 60° C hot plate. The mice in the cage were forced to climb up the side wall of the cage to cool their feet. The result clearly showed that the durability of the skeletal muscle in bestatin treated dystrophic mice had almost completely recovered (Fig. 4).

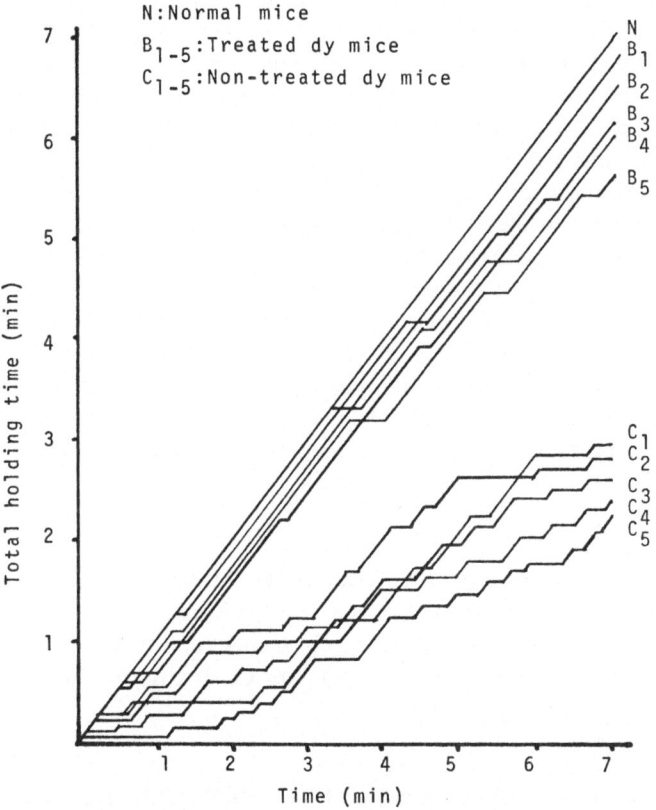

FIG. 4. Endurance test on skeleltal muscle of bestatin-treated and non-treated dystrophic mice.

CONCLUSIONS AND SUMMARY

Histological studies of the nervous system of dystrophic mice have demonstrated that the number of myelinated axon in peripheral nerves is reduced and that hypomyelination exists in the spinal and cranial roots. However, similar studies of the central nervous system of dystrophic mice have not been as extensive.[24]

In the present biochemical study, we have been able to verify the consistent pattern of myelin deficiency existing in the central nervous system of dystrophic mice. Using differential ultracentrifugation, we found a pronounced reduction of the total amount of myelin recovered from the spinal cord of dystrophic mice. Among the reduced concentrations of principal lipid classes, cerebrosides and sulfatides concentrations diminished significantly in the central nervous system of dystrophic mice. It is noteworthy, however, that the total cholesterol level was higher in the spinal cords of dystrophic mice than in the controls. In addition, reduced activities of CNP and CEH were found in purified myelin fraction from the central nervous system of dystrophic mice.

The continuous administration of the protease inhibitors to dystrophic mice at an early stage of the muscle disease resulted in a striking recovery not only of myelin associated enzyme activities (CNP and CEH), but also of the skeletal muscle strength. This suggests that the myelin deficit in the central nervous system might be brought about by an abnormal rise of the catalytic process in the myelin membrane structure, as occurred in the muscle tissue. However, the question of whether the observed abnormalities in the central nervous system of dystrophic mice might contribute, directly or indirectly, to the development of muscular dystrophy remains to be elucidated.

Acknowledgments

The author is indebted to Professor H. Matsushita, Department of Physiology, Wakayama Medical College, for his kind support of this work. This work was supported by research grants for intractable disease and for muscular dystrophy from the Ministry of Health and Welfare, Japan.

REFERENCES

1. Michelson, A. M., Russell, E. S. and Harman, P. J.: Dystrophia muscularis— A hereditary primary myopathy in the house mouse. *Proc. Nat. Acad. Sci. U.S.A.*, **41**: 1079–1084, 1955.

2. Harris, J. G., Wallace, C. and Wing, J.: Myelinated nerve fiber counts in the nerves of normal and dystrophic mouse muscle. *J. Neurol. Sci.*, **15**: 245–249, 1972.

3. Montgomery, A. and Swenarchuk, L.: Further observations on myelinated axon numbers in normal and dystrophic mice. *J. Neurol. Sci.*, **38**: 77–82, 1978.

4. Gilbert, J. J., Steinberg, M. C. and Banker, B. Q.: Ultrastructural alterations of the motor end plate in myotonic dystrophy of the mouse ($dy^{2J}\ dy^{2J}$). *J. Neuropathol. Exp. Neurol.*, **32**: 345–364, 1973.

5. Bradley, W. G. and Jenkinson, M.: Neural abnormalities in the dystrophic mouse. *J. Neurol. Sci.*, **25**: 249–255, 1975.

6. Biscoe, T. J., Caddy, K. W. T., Pallot, D. J. and Pehrson, U. M. M.: Investigation of cranial and other nerves in the mouse with muscular dystrophy. *J. Nuerol. Neurosurg. Psychiat.*, **38**: 391–403, 1975.

7. McComas, A. J. and Mrózek, K.: Denervated muscle fibers in hereditary mouse dystrophy. *J. Neurol. Neurosurg. Psychiat.*, **30**: 526–530, 1967.

8. Lapetina, E. G., Sato, E. F. and De Robertis, E.: Gangliosides and acetylcholinesterase in isolated membranes of the rat brain cortex. *Biochim. Biophys. Acta*, **135**: 33–43, 1967.

9. Folch, J., Lees, M. and Slane-Sanley, G. H.: A simple method for the isolation and purification of total lipids from animal tissues. *J. Biol. Chem.*, **226**: 491–509, 1957.

10. Fiske, C. H. and Subbarow, R. R.: The colorimetric determination of phosphorous. *J. Biol. Chem.*, **66**: 375–400, 1925.

11. Zak, B.: Simple rapid microtechnic for serum total cholesterol. *Amer. J. Clin. Path.*, **27**: 583–588, 1957.

12. Lowry, O. H., Rosebrogh, N. J., Farr, A. L. and Randall, R. J.: Protein measurement with the folin phenol reagent. *J. Biol. Chem.*, **193**: 265–275, 1951.

13. Sorenson, M. and Haugaard, G.: Über die Anwendbarkeit der Orcinreaktion zur Bestimmung der Art und Menge von Kohlenhydratgruppen in Eiweeitzstoffen. *Biochem. Z.*, **260**: 247–277, 1933.

14. Kean, E. L.: Rapid, sensitive spectrophotometric method for quantitative determination of sulfatides. *J. Lipid Res.*, **9**: 319–327, 1968.

15. Lauter, C. J. and Trams, E. G.: A spectrophotometric determination of sphingosine. *J. Lipid Res.*, **3**: 136–138, 1962.

16. Norton, W. T. and Poduslo, S. E.: Myelination in rat brain: Method of myelin isolation. *J. Neurochem.*, **21**: 749–757, 1973.

17. Olafson, R. W., Drummond, G. I. and Lee, J. F.: Studies on 2′,3′-cyclic nucleotide 3′-phosphohydrolase from brain. *Canad. J. Biochem.*, **47**: 961–966, 1969.

18. Sogin, D. C.: 2′,3′-cyclic NADP as a substrate for 2′,3′-cyclic nucleotide 3′-phosphohydrolase. *J. Neurochem.*, **27**: 1333–1337, 1976.

19. Richmond, W.: Preparation and properties of cholesterol oxidase from Nocardia sp. and its application to the enzymatic assay of total cholesterol in serum. *Clin. Chem.*, **19(12)**: 1350–1356, 1973.
20. Kurihara, T., Nassbaum, J. L. and Mandel, P.: 2′,3′-cyclic nucleotide 3′-phosphohydrolase in brains of mutant mice with deficient myelination. *J. Neurochem.*, **17**: 993–997, 1970.
21. Eto, Y. and Suzuki, K.: Enzymes of cholesterol ester metabolism in the brains of mutant mice, quaking and jimpy. *Exptl. Neurol.*, **41**: 222–226, 1973.
22. Aoyagi, T., Takeuchi, T., Matsuzaki, A., Kawamura, K., Kondo, S., Hamada, M., Maeda, K. and Umezawa, H.: Leupeptins, new protease inhibitors from Actinomycetes. *J. Antibiotics*, **22(6)**: 283–288, 1969.
23. Umezawa, H., Aoyagi, T., Suda, H., Hamada, M. and Takeuchi, T.: Bestatin, an inhibitor of aminopeptidase B, produced by Actinomycetes. *J. Antibiotics*, **29(1)**: 97–99, 1976.
24. Bradley, W. G.: Involvement of peripheral and central nerves in murine dystrophy. *Ann. N. Y. Acad. Sci.*, **317**: 132–142, 1979.

Discussion

Dr. Zalc: Is there a crucial period for your treatment to be effective? What is your interpretation of the mechanism of effectiveness of this treatment?

Dr. Tsuji: Yes, there is a critical period for effectiveness of bestatin treatment to prevent disease. I think it is around two weeks of age.

The exact interpretation of the mechanism of effectiveness of this treatment cannot be made under present conditions, but I suppose the first step is suppression of the catabolic inclination of membrane structural protein in muscle cells or nervous tissues of dystrophic mice.

Dr. Ishiura (National Center for Nervous, Mental and Muscular Disorders): Are there any direct findings that these inhibitors suppress endogenous proteinases?

Dr. Tsuji: We have evidence of extreme suppression of CANP activity in the skeletal muscle of dystrophic mice caused by administration of leupeptin or bestatin.

Neurochemical Abnormalities of Snell Dwarf Mutant Mice

Tetsuya Noguchi,* Tetsuro Sugisaki, Ken Takamatsu, and Yasuzo Tsukada

ABSTRACT

In this study, we investigated the neurochemical and histochemical abnormalities of the brains of the Snell dwarf mice. At 40 days of age, among the three parts of the brain (cerebrum, cerebellum, and brain stem), only in the cerebrum were the wet weight, the CNPase activity, and the content of both S-100 protein and Thy-1 antigen found to be biochemically and immunohistologically reduced. In 24-hour records of spontaneous locomotion activity, the dwarf showed a strikingly reduced activity, with an indistinct diurnal rhythm, while the normal controls had a well-marked periodicity. Daily administration of bGH and T_4 to the dwarfs during the first 40 days of postnatal life restored CNPase activity to the level of the normal controls. This was accompanied by normalization of the pattern of spontaneous locomotion activity. Daily administration of bGH alone also restored CNPase activity and spontaneous locomotion, but to a lesser extent. The daily administration to TSH alone, however, failed to restore CNPase activity, in spite of the fact that the thyroid glands of the TSH-treated dwarfs were indistinguishable from the normal controls in terms of organization and appearance.

These results indicate that the restoration of both the retarded myelinogenesis and abnormal behavior of the Snell dwarf mice might essentially depend upon GH levels and the synergistic effects of T_4.

INTRODUCTION

As a consequence of the pituitary being the primary site of the mutant gene's action, the Snell dwarf mice[1] have deficient levels of growth hormone (GH), thyroid stimulating hormone (TSH), prolactin, and possibly corticotrophin.[2-4] The growth of this mutant can be partially restored by administration of pituitary extracts, GH, thyroxine (T_4), prolactin or

Department of Physiology, School of Medicine, Keio University, Tokyo, Japan
* Present address: Second Deparment of Physiology, School of Medicine, Tōhō University, Tokyo, Japan

somatomedin.[2,5,6] In addition, poor myelination and reduced cell number in the brain were also found histologically and biochemically.[7-10]

Several studies have demonstrated that thyroid deficiency in early postnatal life impairs myelinogenesis in the central nervous system.[11-13] Furthermore, Pelton et al.[14] have presented data suggesting that GH deficiency can impair cell migration and oligodendroglial differentiation with subsequent failure of myelin formation. The Snell dwarf mice, therefore, appear to be ideally suited for studying the effects of growth stimulating factors not only on somatic but also on cerebral development.

In this paper, we report on the neurochemical and histological abnormalities of the brains of this mutant and on the effects of bovine GH (bGH), bovine TSH, and T_4 on myelinogenesis. This was investigated by monitoring the activity of the myelin-marker enzyme, 2',3'-cyclic nucleotide 3'-phosphohydrolase (CNPase), and by the level of spontaneous locomotion activity and its diurnal periodicity.

MATERIALS AND METHODS

Reagents
Bovine growth hormone (bGH) was purchased from Miles Laboratories, U.S.A. Bovine thyroid stimulating hormone (TSH) and L-thyroxine (T_4) were purchased from Sigma Co., U.S.A. Other chemicals used were of analytical grade.

Animals
Snell dwarf mice were obtained by mating known heterozygous mice which were purchased from the Jackson Memorial Laboratory, U.S.A. The mice were kept in an air conditioned room with a controlled 12-hour light cycle. Hormones were administered once daily for the first 40 days of postnatal life by subcutaneous injections. The normal control mice (expressed as $+/?$) were used without regard to their actual genotype, since examination of the organs of heterozygous ($dw/+$) or homozygous ($+/+$) normal animals at various ages revealed no statistical difference in total organ weight, contents of DNA, RNA or protein[8] or CNPase activity in the brain.

The normal controls ($+/?$) and dwarfs (dw/dw) were divided into four groups according to treatment given. Daily doses were as follows:
(A) Control groups .... Physiological saline
(B) bGH groups .... 10μg/g body weight
(C) bGH + T_4 groups .... 10μg/g body weight plus T_4
0.25μg/mouse

(D) TSH groups ....0.2 IU of TSH/mouse

CNPase assay

The level of CNPase in the brains was assayed by the method of Kurihara and Tsukada.[15] The reaction product, 2'-AMP, was measured by high-performance liquid microchromatography.[16]

Antisera

The anti-chick myelin basic protein rabbit antiserum and anti-bovine S-100 protein rabbit antiserum were produced in our laboratory. The Theta-C_3H (Thy-1,2) antiserum was purchased from Searle Diagnostic, UK.

Other determination

The content of Thy-1 antigen in the brains was measured by the method of thymus cell cytotoxicity using Theta-C_3H (Thy-1,2) antiserum. The content of S-100 protein was measured by a quantitative complement fixation test using anti-bovine S-100 rabbit antiserum. The determination of contents of serotonin, dopamine, and norepinephrine was done by high-performance liquid chromatography.[17,18]

Spontaneous locomotion activity

Activity was recorded by an Animex Activity Meter Type S (LKB FARAD). The individual patterns of daily locomotion activity of each group were recorded with each column representing activity counts during a 30-min interval.

Immunohistochemical reaction

Mice were killed by vascular perfusion with 10% formaldehyde containing 0.9% NaCl, and the brains were removed and postfixed with the same solution at 4° C. Paraffin-embedded samples were cut into sections 4 μm thick, which were then deparaffinized and equilibrated in Tris-buffered saline. Sections were incubated according to the method described by Sternberger *et al.*[19]

RESULTS AND DISCUSSION

Neurochemical characteristics of the Snell dwarf mice (Table 1)

Weights of the three parts of the dwarf brain at 40 days of age were found to be signicantly lower than those of the normal control mice: 29.3%

TABLE 1. Neurochemical Characteristics of the Snell Dwarf Mice

	Cerebrum		Brain stem		Cerebellum	
	+/?	dw/dw	+/?	dw/dw	+/?	dw/dw
Tissue wet weight (mg)	229±5 (21)	162±3*** (21)	127±5 (21)	98.0±3.5*** (21)	54.2±1.5 (21)	37.7±0.9***(21)
Protein (mg/g wet wt)	131±6 (21)	130±3 (21)	131±4 (21)	131±4 (21)	149±6 (21)	150±7 (21)
CNPase (μmole/min/mg protein)	1.53±0.07 (12)	0.902±0.052*** (12)	4.01±0.17 (12)	3.66±0.16 (12)	1.64±0.07 (12)	1.66±0.13 (12)
Thy-1 antigen (% of +/?)	100	21.5 (2)	100	60.3 (2)	100	42.9 (2)
S-100 protein (μg/g wet wt)	79.9 (2)	45.2 (2)	258 (2)	208 (2)	165 (2)	160 (2)
5-HT (ng/g wet wt)	589 ± 84 (3)	697 ± 49 (3)	915 ± 79 (3)	1018 ± 193 (3)	—	—
DA (ng/g wet wt)	1168 ± 218 (3)	915 ± 104 (3)	1558 ± 312 (3)	1271 ± 185 (3)	—	—
NE (ng/g wet wt)	244 ± 35 (3)	325 ± 73 (3)	758 ± 36 (3)	849 ± 103 (3)	—	—

Results are expressed as mean ± SEM with the number of animals shown in parentheses. Statistical evaluation was by Student's t test, and differences were considered significant at $p < 0.01$. ** and *** indicate $p < 0.005$ and $p < 0.001$, respectively.

lower for the cerebrum, 22.8% for the brain stem, and 30.4% lower for the cerebellum. However, protein contents of the three parts of the *dw/dw* brain were identical to those of the +/?.

CNPase (a marker for myelinogenesis) activity was found to be reduced selectively in the cerebrum of the *dw/dw* (59.0% of the +/?). Contents of Thy-1 antigen (a marker for neuronal surface and synaptogenesis) of the three parts of the *dw/dw* brain were found to be significantly lower than those of the +/?: 21.5% of the +/? cerebrum, 42.9% of the +/? cerebellum, and 60.3% of the +/? brain stem. In addition, the contents of S-100 protein (a marker for glial cells) were found to be reduced selectively in the cerebrum of the *dw/dw* (43.4% of the +/?). The monoamine contents of the *dw/dw* brain was comparable to that of the +/?.

Immunohistochemical observations in the brains of Snell dwarf mice

In order to define, histologically, the biochemical changes of the *dw/dw* brain, localization studies on these markers by the peroxidase-labeled antibody method were carried out on sections of nervous tissues. By using anti-myelin basic protein antiserum (a marker for myelin), the myelin was stained as shown in Fig. 1. In the cingulum and corpus callosum of the *dw/dw* brain (A), there were fewer myelin fibers than in the +/? mice (B). This observation corresponds to the lower CNPase activity in the cerebrum of the *dw/dw*.

Localization studies using anti-Thy-1 antiserum on sections of the cerebrum showed that Thy-1 is associated mainly with gray matter, myelin and the nuclei of neuronal cells were unstained, peroxidase activity was identified not only in the cytoplasm of neurons but also in the axons. These observations are in agreement with the results of Barclay and Hyden.[20] In the *dw/dw* cerebrum, lower activity was identified in the cytoplasm and axons of neurons when compared to those of the +/? (Fig. 2).

In the cerebellum (Fig. 3), the fine granular spots, indicating enzyme activity, were observed in the cytopalsm of the Purkinje cells and in the molecular layers of the +/? mice. The latter was compatible with the presence of Thy-1 on the parallel fibers. In the *dw/dw* cerebellum, there were fewer granular spots in the molecular layers than in that of the +/? mice, although the enzyme activity in the cytoplasm of the Purkinje cells was almost identical to that of the +/? mice. These histological findings are in agreement with the biochemical data mentioned above.

By using anti-S-100 protein antiserum, it was shown that S-100 protein

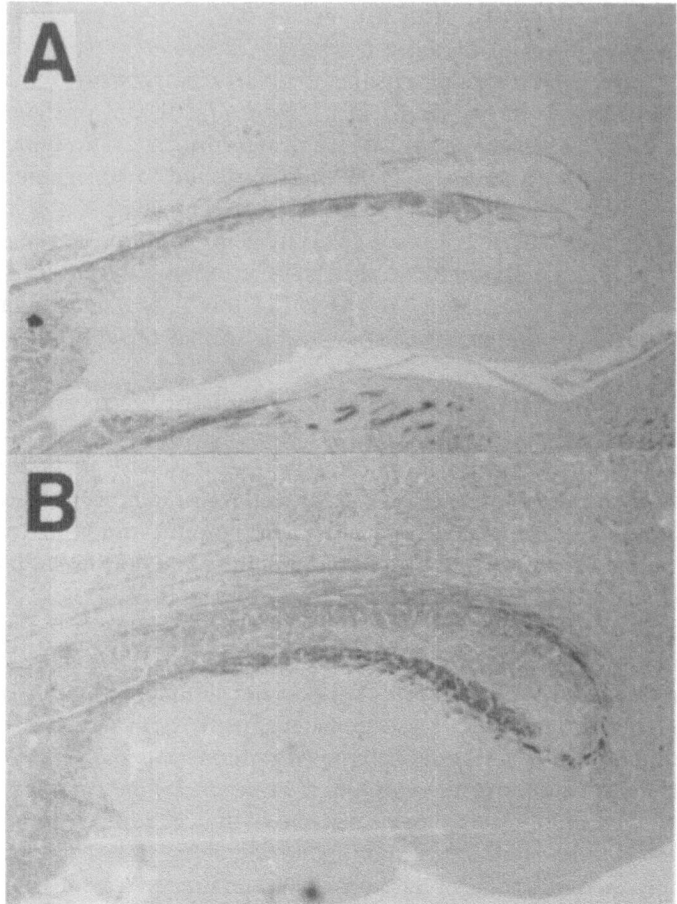

Fig. 1. Immunohistochemical Reaction on the Cerebrum of
the Normal (+/?) and Snell Dwarf (*dw/dw*) Mice with the
Anti Myelin-Basic Protein Antiserum.

(A) *dw/dw* cerebrum: In the cingulum and corpus callosum,
the number of positive myelin-fibers is lower than in those of
the +/? cerebrum.

(B) +/? cerebrum.

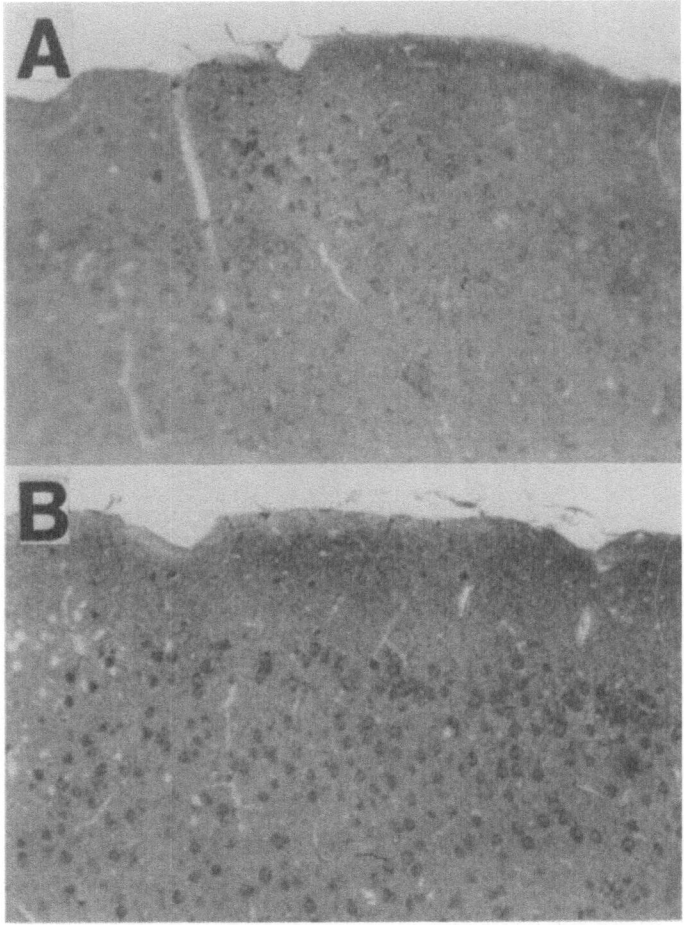

FIG. 2. Immunohistochemical Reaction on the Cerebral
Cortex of the Normal (+/?) and Snell Dwarf (*dw/dw*) Mice
with the Anti Thy-1 Antiserum.

(A) *dw/dw* cerebrum: Lower peroxidase activity is found
in the cytoplasm and axons of neurons when compared to
those of the +/? mouse.

(B) +/? cerebrum: Peroxidase activity is identified not only
in the cytoplasm of neurons but also in the axons. Myelin and
the nuclei of neurons are unstained.

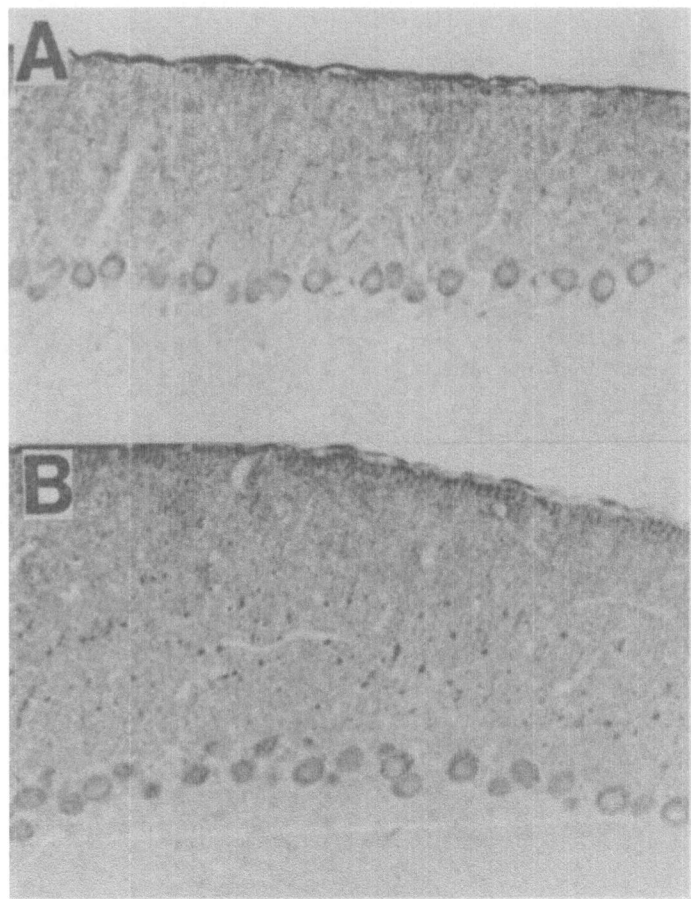

FIG. 3.   Immunohistochemical Reaction on the Cerebellum
of the Normal (+/?) and Snell Dwarf (*dw/dw*) Mice with
the Anti Thy-1 Antiserum.

(A) *dw/dw* cerebellum: Note the smaller number of fine
granular spots of the enzyme activity in the molecular layer.
The cytoplasm of the Purkinje cells is stained as is that of the
+/? cerebellum.

(B) +/? cerebellum.

was mainly localized in the cytoplasm of the glial cells. In contrast, the nuclei of glial cells, myelin, and neuronal cells were unstained. In the cerebral cortex of the *dw/dw* mice, the glial cells were more poorly stained and smaller than in that of the normal controls. This observation is also in agreement with the biochemical data (Fig. 4).

In the cerebellum, both the cell body and fiber of Bergman glia reacted with the antiserum, but no significant difference in the density of S-100 positive cell was observed between the +/? and *dw/dw* mice (Fig. 5).

It is well established that, in the mouse cerebrum, there is a proliferation of neurons prenatally, while glial proliferation occurs postnatally. In the cerebellum, however, glial cells proliferate and migrate postnatally. Therefore, it is understandable that the CNPase activity and the content of S-100 in the cerebellum of the *dw/dw* mice were the same as that of the +/? mice. In the cerebrum, however, as a consequence of impaired glial cell proliferation and maturation, a reduction of CNPase activity and S-100 content might be expected.

Thy-1 antigen is known to localize on the neuronal surface and on the synapse. Maturation of neuronal cells as well as synaptogenesis proceeds postnatally, as does myelination. From the evidence obtained in this experiment showing the lower content of Thy-1 in the *dw/dw* brain, it is likely that neuronal cell maturation and synaptogenesis are also impaired.

The amelioration in the body and brain weights, CNPase activity, thyroid and spontaneous locomotion activity of the Snell dwarf mice after the 40-day replacement therapy

The data obtained in this study indicate that, in the *dw/dw* mice, myelination is still incomplete at 40 days of age, and that maturation of neuronal cells as well as glial cells is arrested during development. In order to determine if GH regulates central nervous system myelination, replacement therapy in the Snell dwarf mice was carried out.

Differences in the body weights of Snell dwarf mice and their normal controls first became apparent on the 10th day of age. The *dw/dw* mice showed no further weight gain until the 40th day of age, while the +/? mice continued to gain weight. In effecting an increase in the body weight of the *dw/dw* mice, the administration of bGH plus T_4 was much more effective than bGH alone. In contrast, TSH failed to cause any increase in body weight. The administration of bGH alone or in combination with T_4 did not result in weight gains for the +/? mice. The TSH-treated +/? mice failed to gain weight, and their body weight on day 40 was significantly

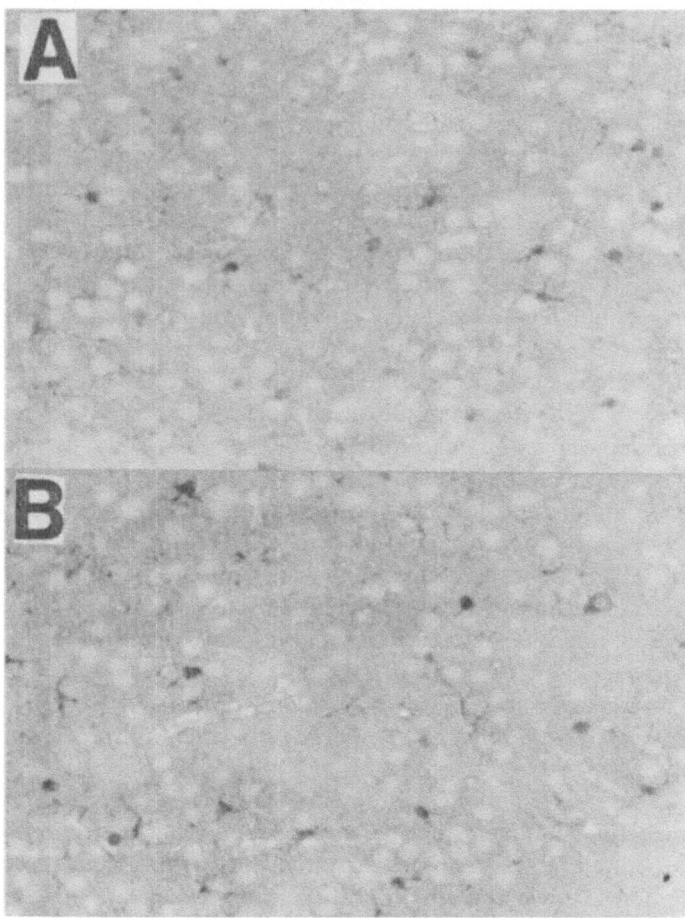

Fɪɢ. 4.   Immunohistochemical Reaction on the Cerebrum of
the Normal (+/?) and Snell Dwarf (*dw/dw*) Mice with the Anti
S-100 Protein Antiserum.

(A) *dw/dw* cerebrum: The glial cells were found to be small
and poorly stained compared to those of the +/? mice.

(B) +/? cerebrum: The cytoplasm of the glial cells is well
stained, whereas the nuclei of the glial cells, myelin, and neuronal cells are unstained.

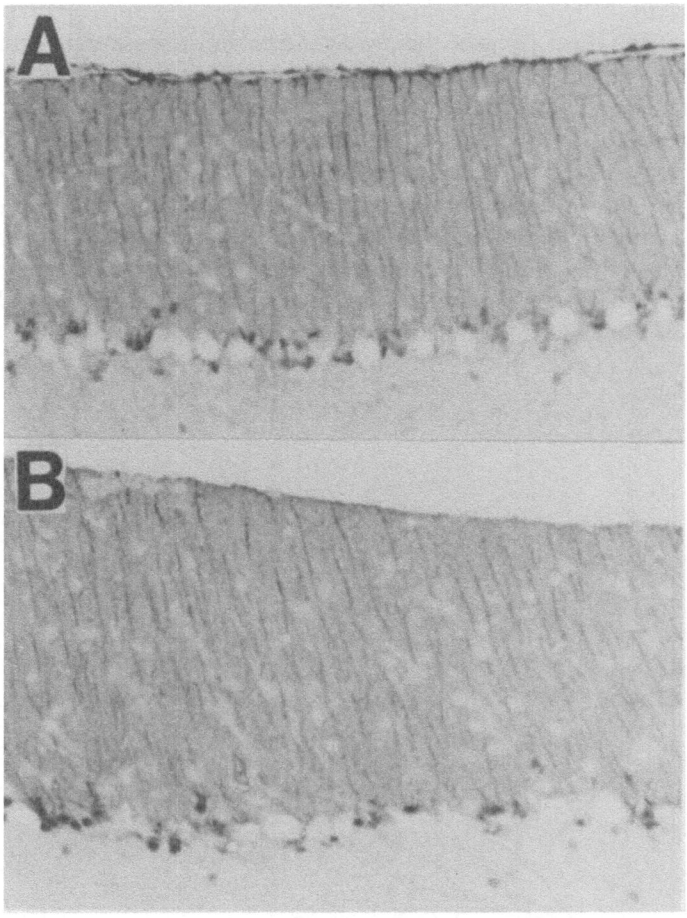

FIG. 5.   Immunohistochemical Reaction on the Cerebellum of the Normal (+/?) and Snell Dwarf (*dw/dw*) Mice with the Anti S-100 Protein Antiserum.

(A) *dw/dw* cerebellum. (B) +/? cerebellum.

Bergman cell bodies and fibers reacted with anti S-100 protein antiserum, but no significant difference in the density of S-100 positive cell was observed between the +/? and *dw/dw* mice.

lower than the $+/?$ mice. This might be the result of hyperfunction of the thyroid (Fig. 6).

Weights of the three parts of the *dw/dw* mouse brain were significantly less than those of the $+/?$ mice. Weights of the cerebrum and cerebellum were increased by the administration of bGH and T_4. When bGH was administered alone, only the weight of the cerebrum increased. TSH administration showed no beneficial effect (Table 2).

CNPase activity was reduced selectively in the cerebrum of the *dw/dw* mice. The administration of bGH and T_4 completely restored it to the normal level. Although the administration of bGH alone tended to increase CNPase activity, the final level was not significantly different from that of the *dw/dw* mice. TSH administration had no noticeable effect on the reduced activity (Table 3).

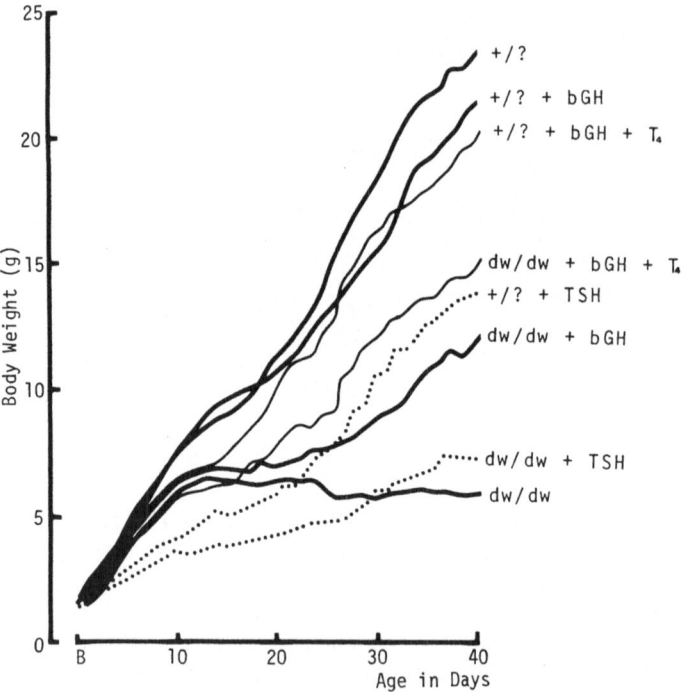

FIG. 6.   Body Weight Changes of Snell Dwarf Mice in Various Hormonal States. Hormonal therapy was the same as in Table 1. Each line represents the mean body weight of ten animals.

TABLE 2. Response of Brain Weights (mg-wet weight) of Snell Dwarf Mice (dw/dw) to Various Hormonal States

Treatments	Cerebrum		Brain Stem		Cerebellum	
	+/?	dw/dw	+/?	dw/dw	+/?	dw/dw
None	229 ± 5 (21)	162 ± 3 (21)	127 ± 5 (21)	98.0 ± 3.5 (21)	54.2 ± 1.5 (21)	37.7 ± 0.9 (21)
bGH	212 ± 18 (6)	184 ± 5 (6)	123 ± 15 (6)	95.0 ± 3.5 (6)	51.2 ± 6.1 (6)	40.0 ± 2.5 (6)
bGH + T$_4$	218 ± 6 (10)	202 ± 7 (10)	118 ± 5 (10)	100 ± 8 (10)	52.1 ± 1.3 (10)	45.1 ± 1.2 (10)
TSH	183 ± 12 (3)	174 ± 12 (7)	104 ± 16 (3)	86.2 ± 3.0 (7)	39.0 ± 5.0 (3)	33.0 ± 5.0 (7)

Hormones were administered daily, by subcutaneous injections, for the first 40 days of postnatal life. Daily doses were as follows: (A) physiological saline only, (B) bGH; 10 μg/g body weight, (C) bGH + T$_4$; 10 μg/g body weight of bGH plus 0.25 μg of T$_4$, and (D) TSH; 0.2 IU per mouse. Results are expressed as mean ± SEM with number of animals shown in parentheses. Statistical evaluation was by Student's t test, and differences were considered significant at $p < 0.01$.

TABLE 3. Response of CNPase Activity (μmoles formed 2'-AMP/min/mg-protein of homogenate) of the Snell Dwarf Mice (dw/dw) to Various Hormonal States

Treatments	Cerebrum		Brain Stem		Cerebellum	
	+/?	dw/dw	+/?	dw/dw	+/?	dw/dw
None	1.53 ± 0.07 (12)	0.902 ± 0.052 (12)	4.01 ± 0.17 (12)	3.66 ± 0.16 (12)	1.64 ± 0.07 (12)	1.66 ± 0.13 (12)
bGH	1.50 ± 0.13 (7)	1.15 ± 0.06 (5)	4.20 ± 0.18 (3)	3.36 ± 0.08 (3)	1.52 ± 0.06 (3)	1.41 ± 0.06 (3)
bGH + T$_4$	1.70 ± 0.03 (13)	1.56 ± 0.04 (15)*	3.33 ± 0.19 (7)	3.29 ± 0.07 (5)	1.48 ± 0.07 (7)	1.57 ± 0.11 (5)
TSH	1.33 ± 0.08 (3)	1.02 ± 0.04 (7)	3.44 ± 0.20 (3)	3.31 ± 0.02 (7)	1.80 ± 0.04 (3)	1.78 ± 0.04 (7)

Hormonal therapy was the same as in Table 1. CNPase was assayed, and the reaction product was measured by high-performance liquid microchromatography. Results are expressed as mean ± SEM with the number of animals shown in parentheses. Statistical evaluation was by Student's t test, and differences were considered significant at $p < 0.01$.

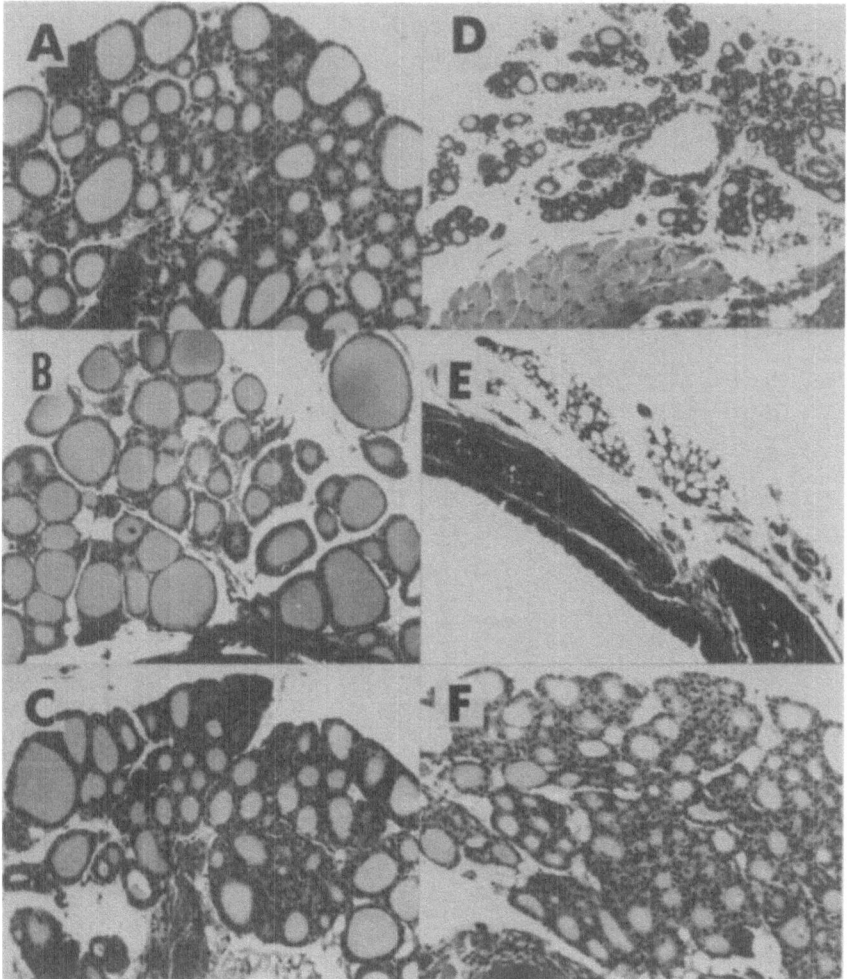

Fig. 7. Comparison of the Thyroids of Normal and Snell Dwarf Mice Receiving Hormonal Therapy during the First 40 Days of Life. Hormonal therapy was the same as in Table 1.

    (A) Normal mouse thyroid.
    (B) The bGH-treated normal mouse thyroid.
    (C) The TSH-treated normal mouse thyroid.
    (D) Untreated dwarf mouse thyroid.   ▶

The thyroids of the *dw/dw* mice were greatly reduced in size. In the center of each lobe, a considerable amount of the tissue was not organized into follicles. TSH administration produced histologically detectable changes in the thyroid; that is, the entire gland became larger and the number of follicles increased at the expense of the central mass of unorganized tissue. The thyroids of the *dw/dw* mice treated with bGH alone, however, showed no histologic response, and were indistinguishable from those of the *dw/dw* mice. Therefore, the possibility that bGH could have been contaminated with TSH has been excluded by this clear-cut difference (Fig. 7).

Figure 8 shows typical 24-hour changes in the spontaneous locomotion activity of the four groups of mice. The *dw/dw* mice exhibited a strikingly reduced level of activity with an indistinct rhythm, while the +/? mice showed spontaneous locomotion activity with a well marked diurnal periodicity. The *dw/dw* mice treated with bGH and T_4 showed a higher level of activity than that of the *dw/dw* mice and maintained a clear diurnal periodicity which was almost identical to the +/?. The *dw/dw* mice treated with bGH alone showed normalization to some extent in both the level of activity and diurnal periodicity.

To summarize what has been mentioned above, dwarf mice receiving only TSH failed to show body and brain weight changes, whereas the administration of bGH alone and in combination with T_4 to the *dw/dw* mice resulted in distinct increases in body and brain weights. The reduced CNPase activity in the *dw/dw* cerebrum rose to the level of the +/? mice following the administration of bGH and T_4 and, to lesser extent, by the administration of bGH alone. TSH administration to the *dw/dw* mice failed to increase either brain weight or CNPase activity, despite complete organization of the immature dwarf thyroid. In addition, the *dw/dw* receiving bGH and T_4 also showed the highest level of spontaneous locomotion activity, with a diurnal periodicity which was virtually identical to that of the normal controls.

Our observations suggest that the acceleration of treated myelinogenesis, possibly through the enhancement of glial cell division, essentially depends upon the GH level plus the synergistic effects of T_4.

◄   (E) The bGH-treated dwarf mouse thyroid. This thyroid is indistinguishable from the untreated dwarf thyroid.

(F) The TSH-treated dwarf mouse thyroid. The entire thyroid is larger and the number of follicles increased at the expense of the central mass of unorganized tissue.

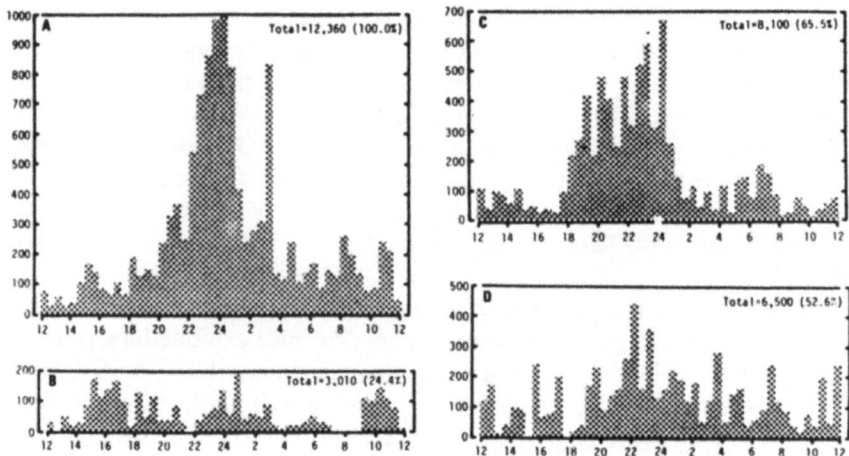

FIG. 8. Effects of Hormones on Spontaneous Locomotion Activity of Snell Dwarf Mice (*dw/dw*). Activity was recorded by an Animex Activity Meter Type S (LKB FARAD). The sensitivity of the activity meter was adjusted to record only locomotion. A single 40-day-old mouse was kept in a plastic box 40 × 30 by 14cm deep placed on the meter. The light cycle was controlled; light was preset to come on at 6:00 and go off at 18:00. Each column represents activity counts during a 30 min interval. The total 24-hour count is shown in the upper right corner of each panel. (A) Normal mice; (B) Untreated Snell dwarf mice; (C) Dwarf mice receiving bGH and T_4; (D) Dwarf mice receiving only bGH.

REFERENCES

1. Snell, G. D.: Dwarf, a new mendelian recessive character of the house mouse. *Proc. Nat. Acad. Sci. U.S.A.*, **15**: 733–734, 1929.
2. Lewis, U. J., Cheever, E. V. and Vanderlaan, W. P.: Alteration of the proteins of the pituitary gland of the rat by estradiol and cortisol. *Endocrinol.*, **76**: 362–368, 1965.
3. Shire, J. G. M. and Hambly, E. A.: The adrenal glands of mice with hereditary pituitary dwarfism. *Acta path. Microbiol. Scand.*, **81**: 225–228, 1973.
4. Sinha, Y. N., Salocks, C. B. and Vanderlaan, W. P.: Pituitary and serum concentration of prolactin and GH in Snell dwarf mice. *Proc. Soc. exp. Biol. (N.Y.)*, **150**: 207–210, 1975.
5. Wallis, M. and Dew, J. A.: The bioassay of growth hormone in Snell's dwarf

mice: Effect of thyroxine and prolactin on the dose-response curve. *J. Endocrinol.*, **56**: 235–243, 1973.

6. Van Buul-Offers, S. and Van den Brande, J. L.: Effect of growth hormone and peptide fractions containing somatomedin activity on growth and cartilage metabolism of Snell dwarf mice. *Acta Endocrinol.*, **92**: 242–257, 1979.

7. Viola-Magni, M.: Cell number deficiencies in the nervous system of dwarf mice. *Anat. Rec.*, **153**: 325–334, 1965.

8. Winick, M. and Grant, P.: Cellular growth in the organs of the hypopituitary dwarf mouse. *Endocrinol.*, **83**: 544–547, 1968.

9. Reier, P. J., Froelich, J. S., Sawchak, J. A. and Hughes, A. F. W.: Maturation of nonmyelinated fiber bundles in a strain of dwarf (Snell's) mice. *Anat. Rec.*, **178**: 103–108, 1974.

10. Reier, P. J., Matthieu, J.-M. and Zimmerman, A. W.: Myelin deficiency in hereditary dwarfism: A biochemical and morphological study. *J. Neuropath. exp. Neurol.*, **XXXIV**: 465–477, 1975.

11. Balazs, R, Brooksbank, B. W. L., Davison, A. N., Eayrs, J. T. and Wilson, D. A.: The effect of neonatal thyroidectomy on myelination in the rat brain. *Brain Res.*, **15**: 219–232, 1969.

12. Walravens, P. and Chase, H. P.: Influence of thyroid on formation of myelin lipids. *J. Neurochem.*, **16**: 1477–1484, 1969.

13. Bass, N. H. and Young, E.: Effects of hypothyroidism on the differentiation neurons and glia in developing rat cerebrum. *J. Neurol. Sci.*, **18**: 155–173, 1973.

14. Pelton, E. W., Young, E., Bass, N. H. and Grindeland, R. E.: Defective myelinogenesis in developing rat cerebrum induced by selective growth hormone deficiency. *Neurology*, **24**: 3777, 1974.

15. Kurihara, T. and Tsukada, Y.: The regional and subcellular distribution of 2',3'-cyclic nucleotide 3'-phosphohydrolase in the central nervous system. *J. Neurochem.*, **14**: 1167–1174, 1967.

16. Tsukada, Y., Nagai, K. and Suda, H.: A rapid micro method for 2',3'-cyclic nucleotide 3'-phosphohydrolase assay using micro high performance liquid chromatography. *J. Neurochem.*, **34**: 1019–1022, 1980.

17. Ponsio, F. and Jonsson, G.: A rapid and simple method for the determination of picogram levels of serotonin in brain tissue using liquid chromatography with electrochemical detection. *J. Neurochem.*, **32**: 129–132, 1979.

18. Felice, L. J., Felice, J. D. and Kissinger, P. T.: Determination of catecholamines in rat brain parts by reverse-phase ion-pair liquid chromatography. *J. Neurochem.*, **31**: 1461–1465, 1978.

19. Sternberger, L. A., Hardy, P. H., Cuculis, J. J. and Meyer, H. G.: The unlabeled antibody enzyme method of immunohistochemistry. Preparation and properties of soluble antigen-antibody complex (horseradish peroxidase-antihorseradish peroxidase) and its use in identification of spirochetes. *J.*

Histochem. Cytochem., **18**: 315–333, 1970.
20. Barclay, A. N. and Hyden, H.: Localization of the Thy-1 antigen in rat brain and spinal cord by immunofluorescene. *J. Neurochem.*, **31**: 1375–1391, 1978.

Discussion

Dr. Mullen: Could you please comment on the reduction of glycerol-3-phosphate dehydrogenase which you mentioned in your abstract? Drs. Marilyn Fisher and Les Lozak, using an antiserum against GPDH, have found it to be reduced in the Bergmann glia but not oligodendrocytes of cerebellar mutants with Purkinje cell disorders.

Dr. Noguchi: The activity of glycerol-3-phosphate dehydrogenease is certainly reduced selectively in the cerebellum of the dwarf mouse. However, in morphological observation using the anti-Thy-1 antiserum, no alteration was observed in the number of Bergmann-glial cells or in their fibers. So at present we have no explanation for the reduced activity of GPDH in the dwarf cerebellum.

Dr. Zalc: Do you think that Thy-1-antigen can really be considered a neuronal marker?

Dr. Tsukada: We assayed the distribution of Thy-1-antigen on the subcellular fractions of the rat brain, and found that it was located most abundantly in the microsomal fraction. In the purified myelin fraction, Thy-1 was found in small amounts. When we tested an amount of Thy-1 on neuronal and glial enriched fractions using the bulk separation technique, Thy-1 content was much higher in the neuronal fraction than in the glial fraction. We thus assumed that Thy-1 was present mainly on the neuronal membrane in the brain.

Characteristics of a New Allele of Tottering Locus, Rolling (*tg*^{rol})

Tatsuo Muroga,* Sen-ichi Oda,** Yoshiro Kameyama,** and Itsuro Sobue*

A new neurological murine mutant called Rolling mouse Nagoya (*tg*^{rol}), which is a new allele of tottering locus, was developed by Oda in 1973.[1] This mutant was found in the course of mating experiments between C57BL/6JNa and SIII, which is characterized by suckling death. Affected mice show a gait disorder which appears to be ataxia. Clinical, genetic anatomopathological, and electrophysiological investigations in addition to histofluorescent analysis were performed in this study in order to determine the causes of this gait disturbance. Histofluorescent analysis has shown abnormal accumulations in the central nervous system, especially in the cerebellum of this mutant, although no remarkable abnormalities were found in the anatomopathological investigations. On the basis of these results, we conducted a pharmacological study to observe the effects on gait disturbance in this mutant mouse.

CLINICAL FEATURES

This mutant has a gait disorder recognized as ataxia, the characteristics of which are creeping or rolling with poor motor control of the hindlimbs. Also, stiffness of the hindlimbs and tail were often observed in this mutant (Fig. 1). This ataxic gait can be first observed between 10 to 14 days after birth. Body hair of this mutant often becomes wetter than that of normal controls, when the moisture content of the air is high. When normal and phenotypically normal mice were mated, the rolling mutation occurred in 26.9% of the offspring. Mating phenotypically normal mice with rolling mice produced 47.6% mutant offspring. Based on these results, we proposed that this condition is transmitted by a single autosomal recessive gene (Table 1).

* First Department of Internal Medicine, School of Medicine, Nagoya University, Nagoya, Japan
** Fourth Division, Research Institute of Environmental Medicine, Nagoya University, Nagoya, Japan.

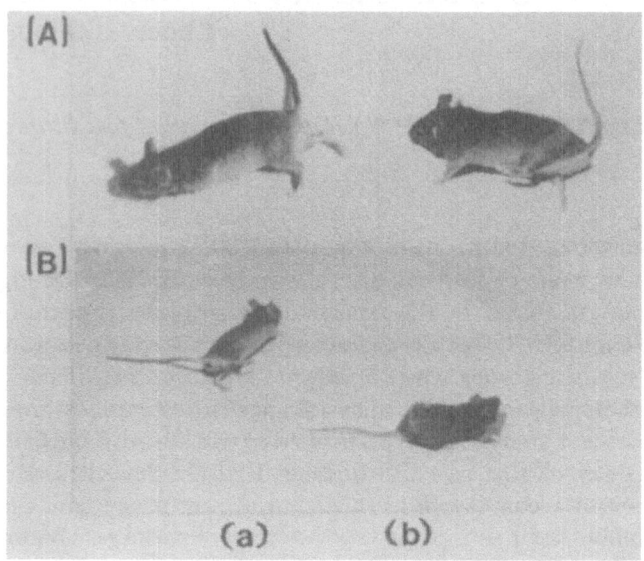

Fig. 1. Abnormal posture of hindlimbs in Rolling mouse
Nagoya.
    [A]: Abnormal posture of hindlimbs in Rolling mouse
        Nagoya.
    [B]: (a) ataxic gait of Rolling mouse Nagoya.
        (b) normal gait of control mouse.

TABLE 1.   Segregation of Rolling (*rol/rol*) in Various Crosses

Matings		No. of observed	No. of rolling mice			No. of normal mice			% rolling
Male ×	Female		♂♂	♀♀	total	♂♂	♀♀	total	
Normal (+/+)	Normal (+/*rol*)	70	0	0	0	30	40	70	0
Normal (+/+)	Rolling (*rol/rol*)	49	0	0	0	24	25	49	0
Normal (+/*rol*)	Normal (+/+)	57	0	0	0	29	28	57	0
Normal (+/*rol*)	Normal (+/*rol*)	201	23	31	54	72	75	147	26.9
Normal (+/*rol*)	Rolling (*rol/rol*)	42	10	10	20	12	10	22	47.6

This table shows *rol* is transmitted by a single autosomal recessive gene.

Anatomopathological investigations performed on this mutant included skeletal muscle and peripherocentral nervous system experiments. No significant defects in cell structure or nerve fiber were found, although Nishida[2] reported that the total number of granular cells in the cerebellum of this mutant decreased to 65% of that of the control mouse. The maximum nerve conduction velocity was measured in five normal control mice and five mutant mice. The normal control group had a conduction velocity of 31.9 m/sec, while the mean of the mutant group was 30.0 m/sec. Therefore, electrophysiologically, there doesn't appear to be a significant difference between these two groups (Fig. 2).

In order to investigate this ataxia further, a tilting board was used to examine the postural response in this mouse. A flat board was attached to the end of the table by hinges and the mouse was fixed on the board with adhesive tapes. Then, needle electrodes were inserted in the extensor of the thigh (Fig. 3). At the same time, changes in the tilting angle of the board

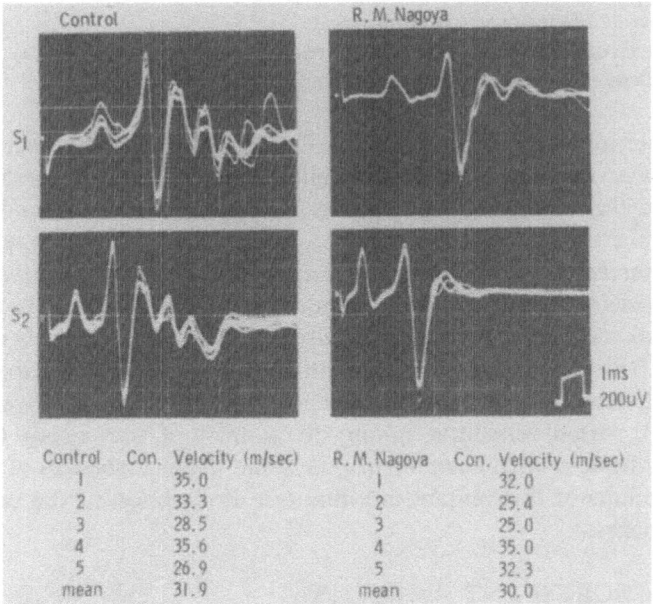

Control	Con. Velocity (m/sec)	R. M. Nagoya	Con. Velocity (m/sec)
1	35.0	1	32.0
2	33.3	2	25.4
3	28.5	3	25.0
4	35.6	4	35.0
5	26.9	5	32.3
mean	31.9	mean	30.0

FIG. 2.   Comparison of maximum nerve conduction velocity of the tail in normal control and Rolling mouse Nagoya.

S_1 and S_2 mean proximal and distal stimulation of the tail, respectively.

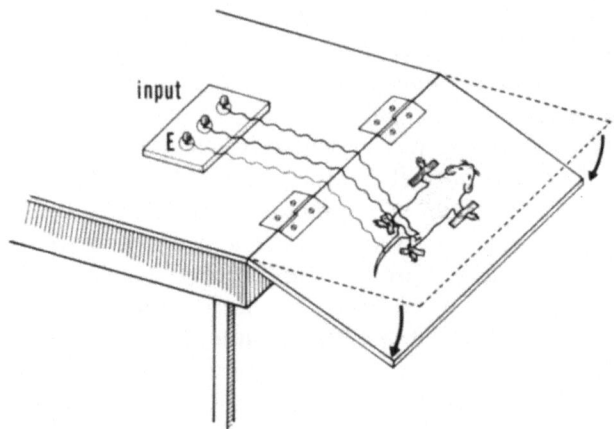

FIG. 3. Experimental instruments for the measurement of evoked responses to tilting load.
Evoked discharges of thigh extensor muscle were repetitively recorded during upward and downward tilting.

were recorded with a potentiometer. The angle of tilt increased and decreased at a constant speed and simultaneous recordings of electromyograms and the tilting angle were done. The latency of the evoked muscle discharge in the normal controls was approximately 20 msec when the tilt was increased in the upward direction. In contrast, the latency of the evoked muscle discharge in the mutant was more than 100 msec (Fig. 4). Downward movement of the board also resulted in significantly different latencies between the groups. Repetitive recordings of the evoked discharges showed that the latency, which occurred with each movement of the board, varied very little among the animals of each group (Fig. 5). Based on these results, it can be proposed that the disturbances of the postural reactions of this mutant are due to a dysfunction of the vestibulospinal reflexes.

HISTOFLUORESCENT ANALYSIS[3]

Ten mutant mice and seven normal littermates weighing 20g each and about 4 months of age were examined. The animals, pretreated with nialamide (25mg/kg), were killed by decapitation 45 minutes after the

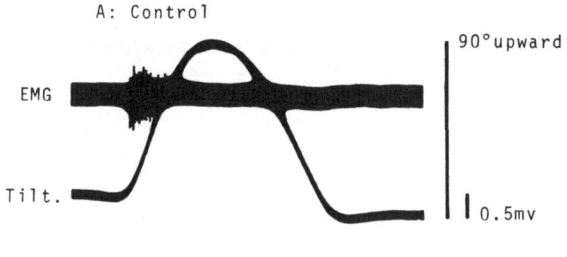

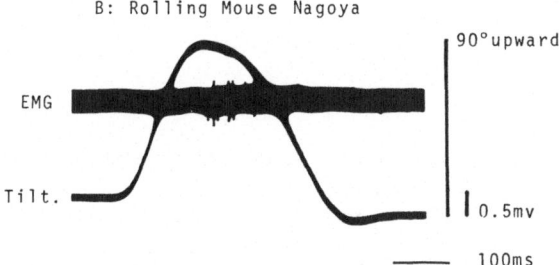

FIG. 4. Examples for evoked discharge of thigh extensor. muscle and tilting angle to tilting load.

treatment and the brains were subjected to histofluorescent analysis by the Falck and Hillarp method.[4-6] In the cerebellum of this mutant, remarkable increases in the number and fluorescent intensity of noradrenaline (NA) nerve terminals were observed throughout all layers, in contrast to those of the littermates. It should be emphasized that there was a topographical difference in the changes in the cerebellum of this mutant. The increases in the number and fluorescent intensity of NA nerve terminals were more distinct in the anterior lobe than in the posterior lobe (Fig. 6). A few NA nerve terminals were present in the cerebellar nuclei of the littermates. In this area of the brain, increases in the number and fluorescent intensity of NA nerve terminals were also observed in the mutants. Similar increases in the number and fluorescent intensity of NA nerve terminals were observed throughout the cerebral cortex of this mutant, compared to the littermates. However, no significant changes in cell number or fluorescent intensity of locus coerulus neurons were found.

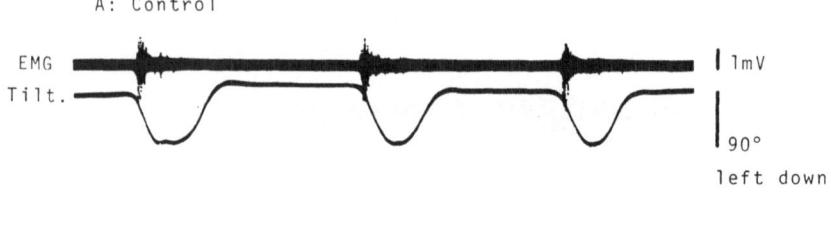

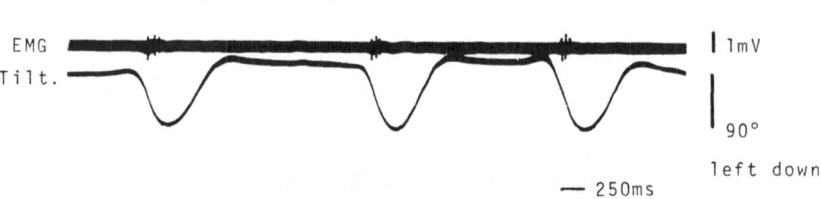

FIG. 5. Successive recordings of evoked responses and alterations of the tilting angle.

    EMG: evoked discharge of thigh extensor muscle.

    Note that the latency of each tilt was almost the same in the animals of each group.

PHARMACOLOGICAL INVESTIGATIONS

Thyrotropin releasing hormone (TRH, 2.5 mg/kg) was intraperitoneally injected and successive observations and recordings of the evoked muscle discharge of the extensor were done. Eight minutes after administration of TRH, unsteadiness of the hindlimbs in this mutant was reduced, gait became easier and an increase in activity was observed. However, thirty minutes after the injection, unsteadiness of the hindlimbs began to reappear. Sixty minutes later, the mouse returned to the pre-injection condition. In addition, TRH injected mice were fixed on the tilting board, and the latencies of the evoked muscle discharge were successively measured. Although there was no change in the latency in the control mice, an increase in spontaneous muscle discharge appeared in the mutants 5 to 15 minutes after the injection. Twenty minutes after the injection, the latency of the evoked discharge was even shorter than in the pre-injection condition (Fig. 7). The latency was 40.2 ± 10.3 msec before the injection and 13.8 ± 5.0 msec 20 minutes after it (Table 2). Reserpine (10 mg/kg) was also intra-

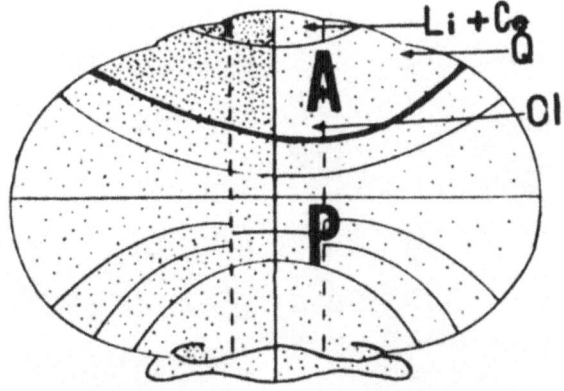

Rolling mouse Nagoya   Control mouse

Fig. 6. Schematic drawing of the topographic difference in the distribution of NA nerve terminals.

Dots indicate NA nerve terminals which are more distinct in the anterior lobe than in the posterior lobe.

A: anterior lobe, P: posterior lobe, Li: lingula, Ce: lobus centralis, Cl: culumen, Q: lobus quadrangularis.

TABLE 2. Changes of Latencies of Evoked Discharge to Tilting Load before and after TRH Injection

	Control	RMN
before	10.1 ± 4.8 (msec)	40.2 ± 10.3 (msec)
5 min.	10.1 ± 2.7	40.2 ± 7.3
10	9.8 ± 3.9	$16.8 \pm 5.8*$
15	9.6 ± 5.6	$16.8 \pm 5.0*$
20	9.8 ± 3.8	$13.8 \pm 5.0*$
25	9.6 ± 5.6	$13.4 \pm 5.9*$
30	9.8 ± 5.8	$16.4 \pm 5.9*$
40	10.2 ± 4.7	19.5 ± 8.8
60	15.8 ± 5.6	23.5 ± 6.3

*: $p < 0.01$

peritoneally injected. Between 10 to 30 minutes after the injection, the latency of this mutant changed from 33.0 ± 3.3 msec to 16.5 ± 1.7 msec. In addition to this change, reserpine decreased the activity of the mice (Table 3).

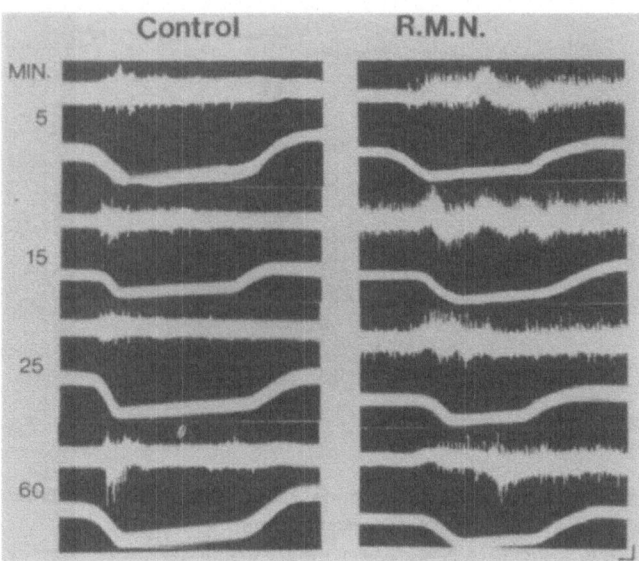

FIG. 7. Examples of changes in latencies of evoked discharge to tilting load after TRH injection.

MIN: minutes, R.M.N.: Rolling mouse Nagoya.

Notice increase of spontaneous muscular discharge and shortening of latency (40 → 13 msec) in Rolling mouse Nagoya. Calibration: 100 μV, 20 msec.

TABLE 3. Changes of Latencies of Evoked Discharge to Tilting Load before and after Reserpine Injection

	Control	RMN
10 min.	11.5 ± 1.0 (msec)	33.0 ± 3.3 (msec)
20	11.6 ± 2.0	18.2 ± 1.7
30	9.9 ± 1.7	16.5 ± 1.7*
40	13.2 ± 1.7	13.2 ± 3.3*
50	9.9 ± 1.0	23.1 ± 3.5
60	9.9 ± 0.7	28.1 ± 5.0

*: $p < 0.01$

DISCUSSION

The cerebellar weights, as well as the body weights, of the mutant mice were lower than those of the controls. However, microscopic investigations did not reveal any remarkable abnormalities. Therefore, we suspect that this mutant may carry metabolic disorders without any morphological murine mutations such as occur in weaver, staggerer and reeler. The abnormal accumulation of NA in the cerebellum has been reported in weaver, staggerer, and reeler mutants.[7] Therefore, it is possible that the same abnormalities might exist in the cerebellar components, including the NA afferent fibers, of Rolling mouse Nagoya. Therefore, the brain of this mutant was studied using a histofluorescent method to examine NA afferent fibers in the cerebellar and cerebral cortices and the locus coeruleus, which is the major source of NA for both cortices. The results suggest that this mutant has high concentrations of NA in the cerebellum as well as in other portions of the brain. It is postulated that TRH facilitates the turnover and consumption of NA metabolism in the cortex of the rat.[8-10] In addition, reserpine plays a role in depleating the storage of NA in the synapses of NA nerve terminals. Therefore, the effects of TRH and reserpine were examined in this mutant. Observations of behavior showed an improvement of ataxic gait, increase of activity and significant shortening of the evoked discharge within 5 to 30 minutes after the injection. Results of reserpine injection also revealed a significant shortening of the latency of the evoked response and a decrease in activity. The TRH effect on motor activity also suggests that NA may be accumulated in the brain of this mutant. Ataxia in this mutant might be due to a disorder of the synapse formation ranging from locus coeruleus to Purkinje cells. In view of these results, we feel that TRH should be studied as a possible treatment for human hereditary spinocerebellar degenerative diseases.

SUMMARY

A new neurological murine mutant, named Rolling mouse Nagoya (tg^{rol}), which is a new allele of tottering locus, was found in the course of mating experiments between SIII and C57BL/6JNa. This mutant shows gait disturbance which appears to be ataxia. The following studies have been performed to investigate the cause of the ataxia.

1. Clinically: Ataxic gait, dysfunction of postural reflex, and autonomic nerve disorders were observed. Postrotatory nystagmus was not observed.

2. Genetically: Autosomal recessive inheritance was determined by the segregation of Rolling (rol/rol) in various crosses.

3. Anatomopathologically: The skeletal muscles, the peripheral and central nervous system did not show any remarkable abnormalities.

4. Electrophysiologically: No significant disturbance of the maximum

conduction velocity was found, although a significant prolongation of latency of the evoked response to tilting load suggested a dysfunction of the vestibulospinal reflexes.

5. Histofluorescent analysis showed a remarkable increase in the number and fluorescent intensity of NA nerve terminals throughout all layers, compared to those of littermates.

6. Pharmacologically: TRH injection in this mouse caused an increase of activity, improvement of ataxic gait, and a significant shortening of the latency of the evoked response to tilting load within 5 to 15 minutes after the injection. Reserpine injection resulted in a significant shortening of the latency and a decrease in activity.

On the basis of these results, we suspect that this mutant mouse carries a metabolic disorder of NA metabolism without any morphological cerebellar abnormalities.

Acknowledgement

We would like to thank Dr. K. Adachi, M.D., for allowing us to use his figure.

REFERENCES

1. Oda, S.: The observation of Rolling mouse Nagoya (*rol*), a new neurological mutant and its maintenance. *Exp. Animals*, **22**(4): 281–288, 1973.
2. Nishimura, Y.: The cerebellum of Rolling mouse Nagoya. *Advances in Neurological Sciences*, **19**(4): 670–672, 1975.
3. Adachi, K., Sobue, I., Toyama, M. and Shimizu, N.: Changes in the cerebellar noradrenaline nerve terminals of the neurological murine mutant, Rolling mouse Nagoya: A histofluorescence analysis. *IRCS Medical Sci.*, **3**: 329–330, 1975.
4. Falck, B., Hillarp, N-Å., Thieme, G. and Torp, A.: Fluorescence of catecholamines and related compounds condensed with formaldehyde. *J. Histochem. Cytochem.*, **10**: 348–354, 1962.
5. Falck, B.: Observation on the possibilities of the cerebellar localization of monoamines by a fluorescence method. *Acta Physiol. Scand.*, **56**: **Suppl. 197**: 1–25, 1962.
6. Carlson, A., Falck, B. and Hillarp, N-Å.: Cellular localization of brain monoamines. *Acta Physiol. Scand.*, **56**: **Suppl. 196**: 1–28, 1962.
7. Landis, S. C., Shoemaker, W. J., Schlumpf, M. and Bloom, F.: Catecholamines in mutant mouse cerebellum fluorescence microscopic and chemical studies. *Brain Res.*, **93**: 253–266, 1975.
8. Keller, H. H., Bartholini, G. and Pletscher, A.: Enhancement of cerebral

noradrenaline turnover by thyrotropin releasing hormone. *Nature*, **248**: 528–529, 1974.

9. Horst, W. C. and Spirt, N.: A possible mechanism for antidepressant activity of thyrotropin releasing hormone. *Life Sci.*, **15**: 1073, 1974.
10. Green, A. R. and Grahame-Smith, D. G.: TRH potentiates behavioral changes following increased brain 5-hydroxytryptamine accumulation in rats. *Nature*, **251**: 524–526, 1977.

Discussion

Dr. Ohno: 1. We obtained a different result regarding the postrotatory nystagmus. We did observe it in the Rolling mouse Nagoya.

2. Is the dysfunction of the vestibulospinal system due to abnormalities in cerebellar function?

Dr. Muroga: 1. We consider this mutant to have wide-spread degrees of motor disturbance, even on the postrotatory nystagmus.

2. I cannot give details, but I think motor disturbance in this mutant is due, not to cerebellipetal abnormality, but probably to dysfunction of the cerebellar (cerebellifugal) efferent pathway, inducing intercortical neuronal disturbance in the cerebellum.

Analysis of CNS Development with Mutant Mice and Chimeras

Richard J. Mullen

Mutant mice provide a wealth of material for using genetics to study the development of the mammalian nervous system. Regardless of whether your prime interest is in the genetics, the development, or the ultimate functioning of the nervous system, the mutants provide a means of "lesioning" this complex system in the hope of revealing a few of the multitude of intricate and complex interactions that are occurring throughout the development and life of the organism. To be certain that the "lesion" is in fact where you think and want it to be, a fundamental question to be asked is, "Where is the primary site of gene action?" To answer this question, it is necessary to allow the "lesioned" cell type access to a normal environment throughout its development and then to observe whether the cell follows its own genetic information or whether its fate is determined by its environment and the cells with which it interacts. To this end we can produce chimeric mice by the relatively simple embryo aggregation technique of Tarkowski[1] and Mintz.[2] By using other genetic variants as independent cell markers, it becomes possible to identify the genotype of a cell regardless of whether or not a cell is expressing the mutant phenotype.

The use of chimeras in analyzing cerebellar mutants has been reviewed recently by Mullen and Herrup.[3] Here I will only attempt to emphasize a few of the main points and to include a few of our more recent results.

THE *pcd* MUTANT

Purkinje cell degeneration (*pcd*) is a relatively new mutant in which all cerebellar Purkinje cells degenerate between 20 and 40 days after birth.[4] In *pcd/pcd* ↔ +/+ chimeras some Purkinje cells do degenerate, the proportion varying between chimeras, indicating that the disorder is not "cured" by the presence of normal cells. To determine the genotype of

Department of Anatomy, University of Utah College of Medicine, Salt Lake City, Utah, U.S.A.

183

the surviving cells, we used the histochemical technique for β-glucuroni-dase developed by Feder.[5] The genotype of the chimera was such that the *pcd* component carried the high-activity allele, *Gus^b*, for β-glucuronidase whereas the normal cells carried the low-activity, *Gus^h*, allele. Thus, the genotype of the chimera was *pcd/pcd Gus^b/Gus^b* ↔ +/+ *Gus^h/Gus^h*. When sections of chimeric cerebellum were stained for glucuronidase, none of the surviving cells were stained, indicating they were +/+ *Gus^h/Gus^h* in genotype. Thus, all of the *pcd/pcd* Purkinje cells had degenerated, indicating that the *pcd* locus acts intrinsically within the Purkinje cell.[6]

These results, however, should not be interpreted to mean that the *pcd* locus acts *only* in Purkinje cells. In fact, we know that the locus acts in many other cell types. As originally reported, *pcd* mutants also lose photoreceptor cells, mitral cells in the olfactory bulb, and males have abnormal or degenerating sperm.[4] This presented a puzzle as to what these diverse cells have in common or whether the *pcd* mutation affected more than one gene. More recently, O'Gorman and Sidman[7] have re-ported cell loss in certain thalamic nuclei including all neurons of the ventral medial geniculate nucleus between postnatal days 40 and 60. Inter-estingly, they also found large accumulations of polysomes similar to those reported by Landis and Mullen[8] in *pcd* Purkinje cells. In the Purkinje cells, however, the basal polysomal mass is a normal developmental feature so that, in the mutants, it appeared to be abnormally retained.

We have recently been examining the mitral cell loss in *pcd* and three interesting observations have emerged. First, the loss of mitral cells is occurring much earlier than we had previously thought.[4] It now appears that there is a massive loss of mitral cells between postanatal days 50 and 70. Secnd, the mitral cells in the accessory olfactory bulb may not be af-fected. Third, the mitral cells also show an accumulation of polysomes (Mullen, unpublished results). Thus, the accumulation of polysomes is a common feature of the degenerations. In Purkinje cells the polysomal mass is abnormally retained during development, whereas in the thalamic neurons and mitral cells it "appears"; this, along with loss of Nissl sub-stance, is known classically as chromatolysis.

PURKINJE CELLS IN OTHER MUTANTS

In the reeler (*rl*), mutant Purkinje cells, as well as most other cortical neurons, are malpositioned. In reeler chimeras we found some Purkinje cells were in normal positions, while others were in positions characteristic

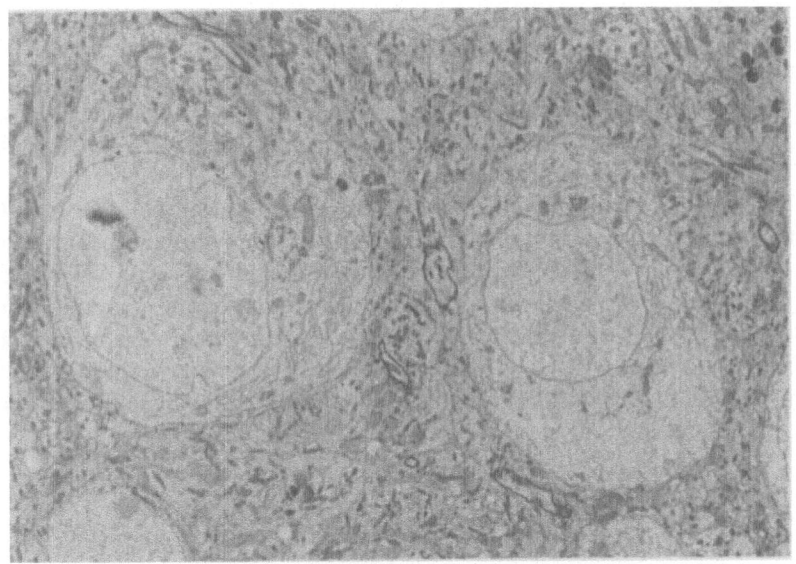

FIG. 1. Two mitral cells from the olfactory bulb of a 40-day-old *pcd/pcd* mutant mouse. In these cells the abnormal accumulation of polyribosomes appears as clear areas at the base of the cells. This chromatolytic change characterizes the degeneration of several cell types in this mutant.

of the reeler. The positioning of the cells, however, was independnt of the genotype of the cell so that gentically normal cells were found in ectopic positions, and, conversely, *rl/rl* cells were found in normal position (Mullen and Sidman, in preparation and 9). Thus, this locus acts extrinsically to the Purkinje cells.

The staggerer (*sg*) mutant is characterized by total degeneration of cerebellar granule cells after they have migrated to the internal granular layer and small ectopic Purkinje cells lacking dendritic spines (the post-synaptic site of the granule cell: Purkinje cell synapse).[10] Because of the latter defect, it was hypothesized that the Purkinje cell might be the site of the defective gene action. Indeed, in staggerer chimeras produced by Dr. Karl Herrup, all of the Purkinje cell defects are expressed, but only by those cells that are *sg/sg* in genotype.[11]

The staggerer chimeras also were a dramatic demonstration of another value of chimeras; that is, they can reveal new aspects of a mutant. For example, in the staggerer chimeras there appeared to be a paucity of Purkinje cells. This led to a reexamination of staggerer mutants, and it was discovered that 60 to 90% of the Purkinje cells were missing.[12] Thus, the observation that the Purkinje cells are lacking dendritic spines, though perhaps very significant in terms of granule cell survival, is somewhat less significant in terms of what the gene is doing considering that the majority of Purkinje cells are entirely absent.

In the nervous (*nr*) mutant, up to 90% of the Purkinje cells degenerate. The feature that characterizes this mutant is the presence of swollen mitochondria in Purkinje cells prior to their degeneration.[13] The surviving Purkinje cells apparently regain normal-appearing mitochondria. Although we have yet to determine wehther the *nr* locus acts intrinsically within the Purkinje cell, we do know that the gene is expressed (at least sometimes) in chimeras, as evidenced by normal and swollen mitochondria

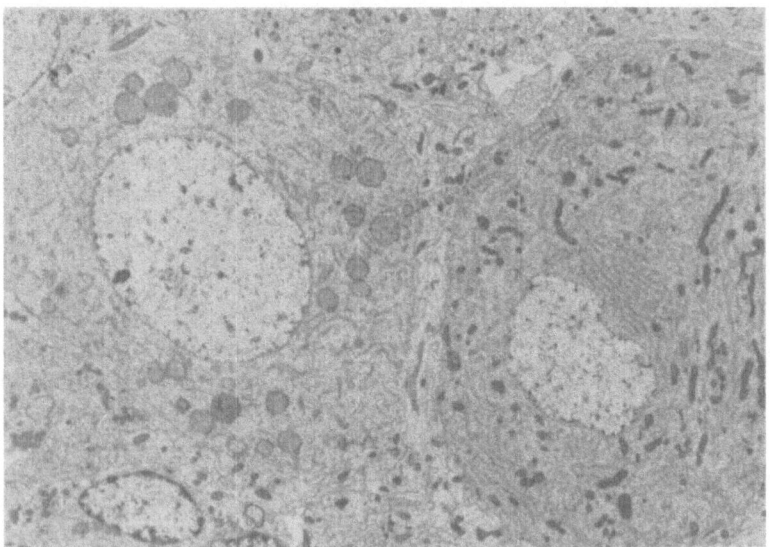

FIG. 2.   Electronmicrograph showing two Purkinje cells from a 21-day-old nervous chimera (*nr/nr* ↔ +/+). The cell on the right appears quite normal while the one on the left shows the swollen mitochondria characteristic of nervous mutants. However, the site of gene action has not been determined yet for this mutant, so we cannot say that the abnormal cell is genetically *nr/nr* nor the other +/+.

being present in adjacent cells. The latter observation must be cautiously interpreted because Sotelo and Triller[14] have observed that in older nr/nr mice degenerating Purkinje cells with the swollen mitochondria can still be found. Our observations, however, were made on young chimeras at times and in regions where all control nr/nr Purkinje cells were exhibiting the swollen mitochondria. Interestingly, nervous mice also exhibit photoreceptor cell degeneration (as do pcd mutants), and the photoreceptors also have the swollen mitochondria.[15]

THE WEAVER MUTANT

Most recently Dr. Dan Goldowitz and I have been studying the weaver (wv) mutant. In homozygous wv/wv mutants, the cerebellar granule cells fail to migrate and degenerate in the external granule cell layer, but there has been controversy as to whether the defect is in the granule cell[16] or the Bergmann glia,[17] which act as guides for the migrating granule cells. Rezai and Yoon[18] discovered that heterozygous $+/wv$ were also affected. The $+/wv$ are similar to the wv/wv only the effects are less severe. In $+/wv$, many granule cells fail to complete their migration to the internal granular layer and remain in ectopic positions in the molecular layer. In addition, the Purkinje cells are also ectopic, not being aligned in a single cell lamina as they are in $+/+$ mice.

We have produced $+/wv$ $Gus^b/Gus^b \leftrightarrow +/+$ Gus^h/Gus^h chimeras to determine if the Purkinje cell ectopia is a primary effect of the gene. In the chimeras we did find numerous ectopic Purkinje cells, and with the glucuronidase marker found that the ectopic Purkinje cells were both mutant and normal in genotype.[19] Thus, the Purkinje cell ectopia appears to be secondary to some other defect caused by the wv locus.

To determine whether the wv locus was acting within the granule cell, a new cell marking system had to be developed. Since the granule cell has very little cytoplasm, we turned to a mutant with a nuclear abnormality. It had earlier been reported by Green et al.[20] that ichthyosis (ic) mutant mice (so named because of their very sparse hair) had abnormally clumped heterochromatin in leucocytes and a few other cell types. We have examined the CNS of ichthyosis mutants and have found a similar nuclear abnormality in sooe cell types, including cerebellar granule cells. In $+/+$ granule cells, the heterochromatin is usually present as two or three small clumps along the nuclear membrane. In ic/ic it is much more common to find a single, large, centrally placed mass of heterochromatin.[21] There is, however, some overlap in phenotypes. For example, in a single 1 μm

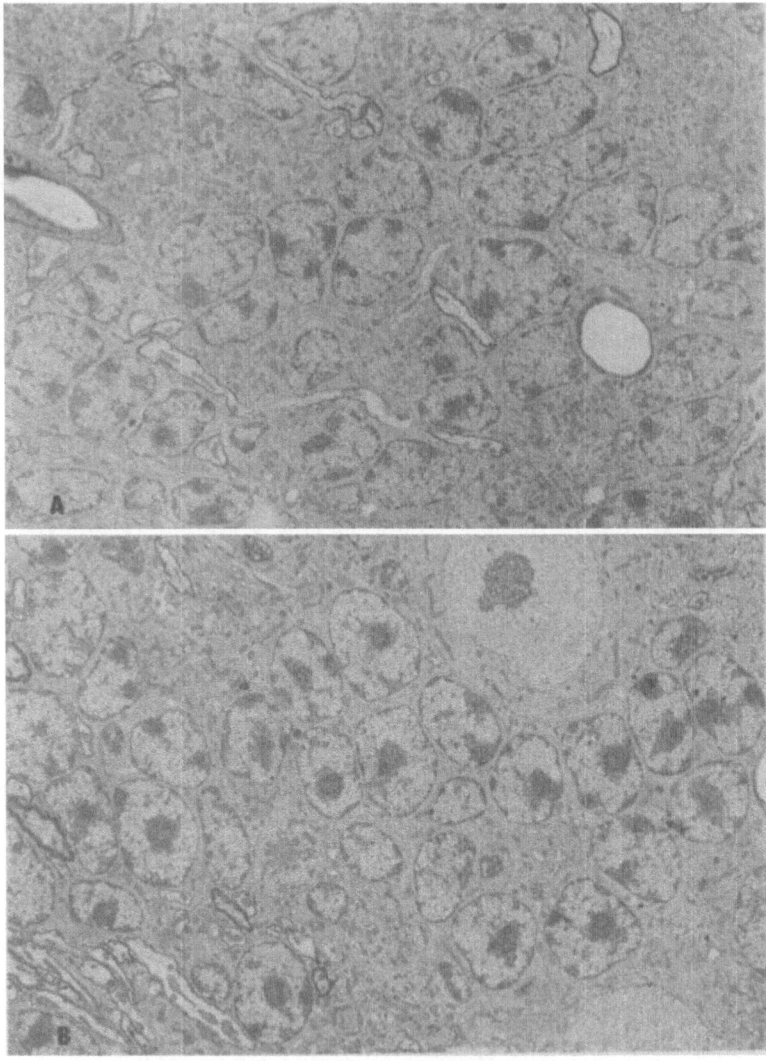

FIG. 3.   Electromicrographs of cerebellar granule cells. (A)
+ / granule cells showing the normal appe arance of heterochromatin distributed as small clumps, usually along the
nuclear membrane. (B) *ic/ic* granule cells, many showing the
large, centrally placed mass of heterochromatin.

plastic section of $+/+$ cerebellum, about 5% of the granule cells will exhibit a centrally placed mass of heterochromatin (i.e., the IC phenotype) whereas in a similar section of ic/ic cerebellum, 25 to 30% will exhibit the IC phenotype (although serial reconstruction showed that all ic/ic cells had the clumped heterochromatin). Thus, at present, the ichthyosis marker is most useful as a quantitative marker for determining the composition of a population of cells rather than for determining unequivocally the genotype of single cells. In the analysis of $+/wv \leftrightarrow +/+$ chimeras that follows, the analysis must be quantitative anyway since even in normal $+/+$ mice one finds ectopic granule cells.

Dr. Goldowitz and I[19] have examined several $+/wv$ $+/+ \leftrightarrow +/+$ ic/ic chimeras and found that the number of ectopic granule cells in the chimeras was intermediate between $+/wv$ and $+/+$ controls. Using the ichthyosis cell marker we found that in the internal granular layer there was a mixture of $+/wv$ and $+/+$ cells as would be expected since granule cells do survive in heterozygous mutants. When we scored ectopic granule cells in the molecular layer, however, we found them to be overwhelmingly $+/wv$ in genotype. This observation that apparently only the $+/wv$ granule cells are failing to migrate normally suggests that the wv locus is acting in the granule cell. If the wv locus were acting in the Bergmann glia then $+/+$ granule cells would presumably also have difficulty migrating. However, we have not ruled out the possibility that both cell types are directly affected by the wv locus. For example, it is possible that in the chimeras we have examined there were sufficient $+/+$ normal Bergmann glia to allow the $+/+$ granule cells to migrate. The $+/wv$ granule cells, being directly affected by the gene, would still, however, have difficulty migrating normally.

MOSAIC PATTERN

Another principal use of chimeras has been in studies attempting to elucidate the distribution of cells for evidence of "clones". Reviews of this use of chimeras can be found elsewhere.[22,23] We have used two approaches in attempting to find evidence of clones. One approach was to use the pcd chimeras described above. As noted, all of the mutant cells degenerated in these chimeras so we were left with only the genetically normal cells. In pcd mutants, the Purkinje cells degenerate after the cortex has been formed so it would seem unlikely that there would be any extensive rearrangement of the $+/+$ cells after the pcd/pcd cells degenerated. A portion of the cerebellar cortex of one $pcd/pcd \leftrightarrow +/+$ chimera in which 75% of the Purkinje cells had degenerated was serially reconstructed to show the distribution of the surviving $+/+$ cells.[6] The visual impression was that there were no discrete patches other than what one would expect in a random array.

The other approach we used was to analyze $Gus^b/Gus^b \leftrightarrow Gus^h/Gus^h$ chimeras stained for β-glucuronidase to compare the observed patch size (i.e., cells of like genotype) with the patch size expected in a random array. From such a comparison an estimate of clone size can be made. Based on an analysis of nearly 2,000 cells in 270 patches, we found that the number of cells observed in a patch was nearly identical to what would be expected in a random array.[6] Subsequently, we found that the proportion of cells of like genotype varied between different cerebellar regions, indicating that the Purkinje cells were not randomly distributed throughout the entire cerebellum.[22] Thus, there was some evidence of clonal development but within small regions of the cortex the cells were extensively intermingled. We were unable to answer the intriguing question as to when during development the intermingling occurred, because the β-glucuronidase marker is not suitable for use on embryos and in addition, it is extremely difficult to distinguish Purkinje cells from other cells during the embryonic and perinatal periods.

Results of a similar type of study have recently been reported by Oster-Granite and Gearhart.[24] They used a new cell marking system which they have developed consisting of an antibody against an electrophoretic variant of glucosephosphate isomerase ($Gpi-1^b$) visualized on tissue sections by immunofluorescence.[25, 26] Their results were similar to ours in that they found regional variation within the cerebellum and that the Purkinje cells of like genotype existed in small patches. However, there was a notable difference in that they did not find the cells to be randomly arranged but rather in patches consisting of 3 to 8 cells, which is larger than would be expected in a linear random array. Although they used the same statistical analysis as us, there are some differences between the studies. First, different cell marking systems were used. I am convinced of the validity of the glucuronidase marking system for Purkinje cells and their system seems equally valid, so I doubt that this is a factor. Second, they did serial reconstructions of three sections for a total of 15 μm. Reconstruction is a potential source of error and the 15 μm is about twice as thick as our material. However, I do not see how either of these could give rise to differences of any magnitude. Finally, we used different strain combinations and this could be the key to our differences, for West[23] has shown that in other systems, such as retinal pigment epithelium, you can get different results with different strain combinations, and even Oster-Granite and Gearhart[24] obtained different estimates of patch size from the different chimeras. Thus, there is a very good chance the differences are real.

CONCLUSION

The above studies hopefully give some idea of the usefulness of chimeras in developmental neurobiology. The studies can often be quite time-consuming and the results of many months of work can often be summarized in a single sentence. Nevertheless, the reward is that chimeras are exquisite systems for studying cell interactions and allow us to answer questions about site of gene action that would be extremely difficult, if not impossible, to answer otherwise.

At one time, inadequate cell markers were hampering the use of chimeras. Now with β-glucuronidase, ichthyosis, anti-GPI antibodies, etc., the opportunity exists for fully exploiting chimera technology as just one of many new tools in developmental neurobiology.

Acknowledgement

I am indebted to Drs. Oster-Granite and Gearhart for allowing me to see preprints of their paper on cell lineage.

This research was supported by the National Institutes of Health (NS 16156) and by the March of Dimes Birth Defects Foundation Basic Research Grant #1–684.

REFERENCES

1. Tarkowski, A. K.: Mouse chimaeras developed from fused eggs. *Nature*, **190**: 857–860, 1961.
2. Mintz, B.: Genetic mosaicism in adult mice of quadriparental lineage. *Science*, **148**: 1232–1233, 1965.
3. Mullen, R. J. and Herrup, K.: Chimeric analysis of mouse cerebellar mutants. In: Neurogenetics: Genetic Approaches to the Nervous System (ed. X. O. Breakefield), pp. 173–196. Elsevier, New York, 1979.
4. Mullen, R. J., Eicher, E. M. and Sidman, R. L.: Purkinje cell degeneration, a new neurological mutation in the mouse. *Proc. Nat. Acad. Sci. U.S.A.*, **73**: 208–212, 1976.
5. Feder, N.: Solitary cells and enzyme exchange in tetraparental mice. *Nature*, **263**: 67–69, 1976.

6. Mullen, R. J.: Site of *pcd* gene action and Purkinje cell mosaicism in cerebella of chimeric mice. *Nature*, **270**: 245–247, 1977.

7. O'Gorman, S. V. and Sidman, R. L.: Cell loss in diencephalic nuclei of Purkinje cell degeneration (*pcd*) mutant mice. *Soc. Neurosci. Abstr.*, **6**: 81, 1980.

8. Landis, S. C. and Mullen, R. J.: The development and degeneration of Purkinje cells in *pcd* mutant mice. *J. Comp. Neurol.*, **177**: 125–143, 1978.

9. Mullen, R. J.: Genetic dissection of the CNS with mutant-normal mouse and rat chimeras. In: Society for Neuroscience Symposia, Vol. 2 (ed. W. M. Cowan and J. A. Ferrendelli), pp. 47–65. Society for Neuroscience, Bethesda, Maryland, 1977.

10. Landis, D. M. D. and Sidman, R. L.: Electron microscopic analysis of postnatal histogenesis in the cerebellar cortex of staggerer mutant mice. *J. Comp. Neurol.*, **179**: 831–863, 1978.

11. Herrup, K. and Mullen, R. J.: Staggerer chimeras: Intrinsic nature of Purkinje cell defects and implications for normal cerebellar development. *Brain Res.*, **178**: 443–457, 1979.

12. Herrup, K. and Mullen, R. J.: Regional variation and absence of large neurons in the cerebellum of the staggerer mouse. *Brain Res.*, **172**: 1–12, 1979.

13. Landis, S.: Ultrastructural changes in the mitochondria of cerebellar Purkinje cells of nervous mutant mice. *J. Cell Biol.*, **57**: 782–797, 1973.

14. Sotelo, C. and Triller, A.: Fate of presynaptic afferents to Prukinje cells in adult nervous mutant mouse: A model to study presynaptic stabilization. *Brain Res.*, **175**: 11–36, 1979.

15. Mullen, R. J. and LaVail, M. M.: Two new types of retinal degeneration in cerebellar mutant mice. *Nature*, **258**: 528–530, 1975.

16. Sotelo, C. and Changeux, J. P.: Bergmann fibers and granular cell migration in the cerebellum of homozygous weaver mutant mouse. *Brain Res.*, **77**: 484–491, 1974.

17. Rakic, P. and Sidman, R. L.: Weaver mutant mouse cerebellum: Defective neuronal migration secondary to specific abnormality of Bergmann glia. *Proc. Nat. Acad. Sci. U.S.A.*, **70**: 240–244, 1973.

18. Rezai, Z. and Yoon, C. H.: Abnormal rate of granule cell migration in the cerebellum of "weaver" mutant mice. *Devel. Biol.*, **29**: 17–26, 1972.

19. Goldowitz, D. and Mullen R. J.: Granule cell as a site of gene action in the weaver mouse cerebellum: Evidence from heterozygous mutant chimeras. *J. Neuroscience* (in press).

20. Green, M. C., Shultz, L. D. and Nedzi, L. A.: Abnormal nuclear morphology of leukocytes in the mouse mutant ichthyosis, a possible transplantation marker. *Transplantation*, **20**: 172–175, 1975.

21. Goldowitz, D. and Mullen, R. J.: Nuclear morphology of ichthyosis mutant mice as a cell marker in chimeric brain. *Devel. Biol.*, **89**: 261–267, 1982.

22. Mullen, R. J.: Mosaicism in the central nervous system of mouse chimeras. In: The Clonal Basis of Development, 36th Symposium of the Society for Developmental Biolology, pp. 83–101. Academic Press, Inc., New York, 1978.

23. West, J. D.: Clonal development of the retinal epithelium in mouse chimaeras and X-inactivation mosaics. *J. Embryol. exp. Morph.*, **35**: 445–461, 1976.

24. Oster-Granite, M. L. and Gearhart, J.: Cell lineage analysis of cerebellar Purkinje cells in mouse chimeras. *Devel. Biol.*, **85**: 199–208, 1981.

25. Gearhart, J. and Oster-Granite, M. L.: An immunofluorescence procedure for the tissue localization of glucosephosphate isomerase. *J. Histochem. Cytochem.*, **28**: 245–249, 1980.

26. Oster-Granite, M. L. and Gearhart, J.: Immunofluorescence and histochemical localization of glucosephosphate isomerase in neural tissue. *J. Histochem. Cytochem.*, **28**: 250–254, 1980.

Discussion

Dr. Fukuda: Is it possible that the presence of β-glucuronidase itself in P-cells, not the genes which induce abnormalities, modifies or even disturbs the arrangement of nerve cells in the chimeric mice?

Dr. Mullen: No, I do not think that is a possibility. The β-glucuronidase is an independent locus that has nothing to do with the neurological mutants. The glucuronidase variants are perfectly healthy mice.

Immunohistochemical and Biochemical Analyses of Development of Nervous System of Mutant Mice (Reeler and Shiverer)

Katsuhiko Mikoshiba,* Ken Takamatsu,* Shin-ichi Kohsaka,*
Yasuzo Tsukada,* and Yoshiro Inoue**

ABSTRACT

Among the various neuropathological mutant mice, reeler and shiverer, which have mutations in the different steps of neurogenesis were chosen for this study.

In the reeler cerebrum, the first formed cells occupy the most superficial part and the last formed cells occupy the deepest part, whereas in the controls the first formed cells stay at the deepest part of the cortex. Reeler is characterized by an inversion of the cortical layers of the cerebrum. Reeler cerebellum is also characterized by an inversion of granular and Purkinje cells. The pathogenesis of the reeler mutation is considered to be an inability of cells to migrate after proliferation. It is therefore a good model to study what factor controls cell migration in the central nervous system (CNS). By immunohistochemical staining of the myelinated fibers using antiserum against myelin basic protein (MBP), it was suggested that the relative position of neurons is not critical to the morphogenetic mechanism that controls the intercellular connections during development, namely the cells must have known their specific targets in terms of the mutual contact of cells. By the 2-DG technique, it was found that altered fiber connection might work functionally even in an inverted cell position.

Development of the reeler cerebellum is greatly affected by the mutation. Since most of the cells divide after birth in the cerebellum, it is possible that the inversion of layers might exert some damage on the dividing cells. In the primary culture system, where cellular contact is relatively easy, the development of the reeler cerebellum was normal in spite of an abnormal Purkinje cell position. The normal synaptic contact, therefore, might be an important factor for the development of the brain. Immunohistochemical staining of the myelinated fibers and the 2-DG technique suggest that there is a reorganization of the cerebellar circuitry in the reeler.

Shiverer mouse is characterized by poor lamella formation and absence of

* Department of Physiology, School of Medicine, Keio University, Tokyo, Japan
** Deparment of Anatomy, School of Medicine, Hokkaido University, Sapporo, Japan

195

a major dense line in the CNS myelin. The immunohistochemical technique demonstrated the absence of MBP in the shiverer. SDS-polyacrylamide gel electrophoresis revealed that the small and large MBP and intermediate protein are missing and that PLP was relatively decreased and Wolfgram protein relatively increased. The activities of 2′,3′-cyclic nucleotide 3′-phosphohydrolase (CNPase), myelin marker enzyme, did not show a significant difference between the shiverer and the control, although there is a morphological dysmyelination in the shiverer. Analysis of the myelin from the peripheral nervous system (PNS) by SDS-polyacrylamide gel electrophoresis showed that MBP common to that of the CNS were missing, whereas P0 and P2 proteins specific to PNS were found to be almost at the same level as the control. Even in the absence of myelin components, the lamella formation was well developed in the PNS. In contrast, dysmyelination was clearly observed to accompany the absence of the same myelin components. It was also found by electron microscopy that the recognition step of axons by the oligodendrocytes was normal but the next step of wrapping the axon was abnormal. The shiverer offers a good model for the study of the mechanism of myelinogenesis.

INTRODUCTION

Neurogenesis is a long process which begins from cell proliferation at germinal zones located at characteristic places in the developing nervous system.[1] Various types of neurons and glia are generated from the homogeneous population of germinal cells, by which the number of cells generated are determined. The next step is cell migration to reach a correct position in the nervous system, and the final step is cell differentiation.

Recently, many neuropathological mutant mice have been reported that are affected in the various steps of neurogenesis.[2-4] Two mutant mice, reeler and shiverer, will be described in this article. The reeler mutant is affected in the cell migration step. Cell migration is one of the important processes peculiar to the vertebrate nervous system. Another mutant, shiverer, is affected in the myelination step, which is directly related to the functional activity of the brain. Analyses of these two mutants would contribute to an understanding of the mammalian nervous system.

REELER MUTANT MOUSE

The reeler is an autosomal recessive mutation found in mice.[5] Homozygous reeler mice have action tremor, dystonic posture, and reeling ataxic gait. There are abnormalities of the cerebral and cerebellar cortices and hippocampus, which are characterized by malpositon of neurons.[3,6-13]

Lissencephaly is a human malformation which is characterized by the absence of convolutions of the cerebral hemisphere.[14] There is extensive subcortical heterotopic masses of neurons arrested in migration. Reeler, therefore, is a condition similar to lissencephaly in humans.

In the normal cerebrum, the first formed cells migrate after proliferation and occupy the deepest part of the cerebrum. The second formed cells migrate through the first formed layer and occupy the upper region of the cerebrum. The last formed cells migrate through the layers already formed to form a layer very near the surface of the cortex (Fig. 1).

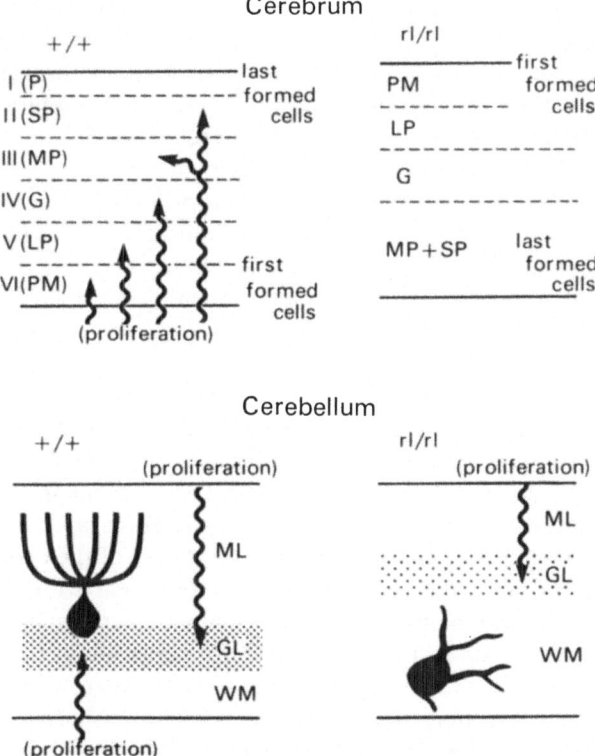

FIG. 1. Scheme of the formation of cortical layers in the cerebrum and the cerebellum from control and reeler mice.

+/+, control mice; rl/rl, reeler mice; P, plexiform layer; SP, small pyramidal layer; MP, medium pyramidal layer; G, granule cell layer; LP, large pyramidal layer; PM, polymorphic cell layer; ML, molecular layer; GL, granular layer; WM, white matter.

Reeler, however, is characterized by the inability of cells to migrate. The second formed cells push up the first formed cells which stay at the upper part and the last formed cells push up all the existing cell layers. The result is that the first formed cells always occupy the most superficial part and the last formed cells the deepest part of the cortex. In other words, the reeler mutant is characterized by inverted cortical layers in the cerebrum (Fig. 1).

There is an inversion of layers in the reeler cerebellum as well. In the control cerebellum, Purkinje cells first migrate to form a cell layer and then granule cells which proliferated postnatally migrate to form a granular layer below the Purkinje cell layer. In the reeler cerebellum, the Purkinje cells cannot migrate to form a cell layer. Therefore, they are localized in the white matter below the granule cell layer with some of the cells localized in the granular layer or in a normal position (Fig. 1). No migration abnormlities are observed in the granule cells in the reeler.[13]

The abnormalities in cortical cell arrangement associated with the reeler mutation pose three main questions of interest in developmental biology. First, what is the factor that controls the position of neurons in the cortical laminar alignment of the brain? Second, what is the genetically altered mechanism that governs connections between cells? Third, how is the function of the brain changed by its genetically altered connections?

Development of the cerebral cortex of the reeler mutant mice

The sequential process of formation of the cerebral cortex during postnatal development is shown in Figs. 2a–c. The cells which proliferated at the germinal zone at the roof of the lateral ventricle migrate immediately up to the surface of the cortex sparing a plexiform layer in the control mice. In the reeler, the cells stay at the central part of the cerebral cortex. On the third postnatal day, most of the cells disappear from the germinal zone near the lateral ventricle of the cerebrum of both reeler and control following the same time schedule (Fig. 2a). The gradient of the cell population in reeler is just the opposite of that of the control (Fig. 2a). On the ninth postnatal day, there is still a cellular gradient in both reeler and the control but the distribution becomes much more even than in the earlier stages (Fig. 2b). In the adult stage, cortical layers are composed of evenly

reeler cortex, which do not migrate, stay at the central part ▶ of the brain, resulting in a clear difference of cell populations in the reeler and the control. (Mikoshiba *et al.*, *J. Neurochem.*, 1980)[12]

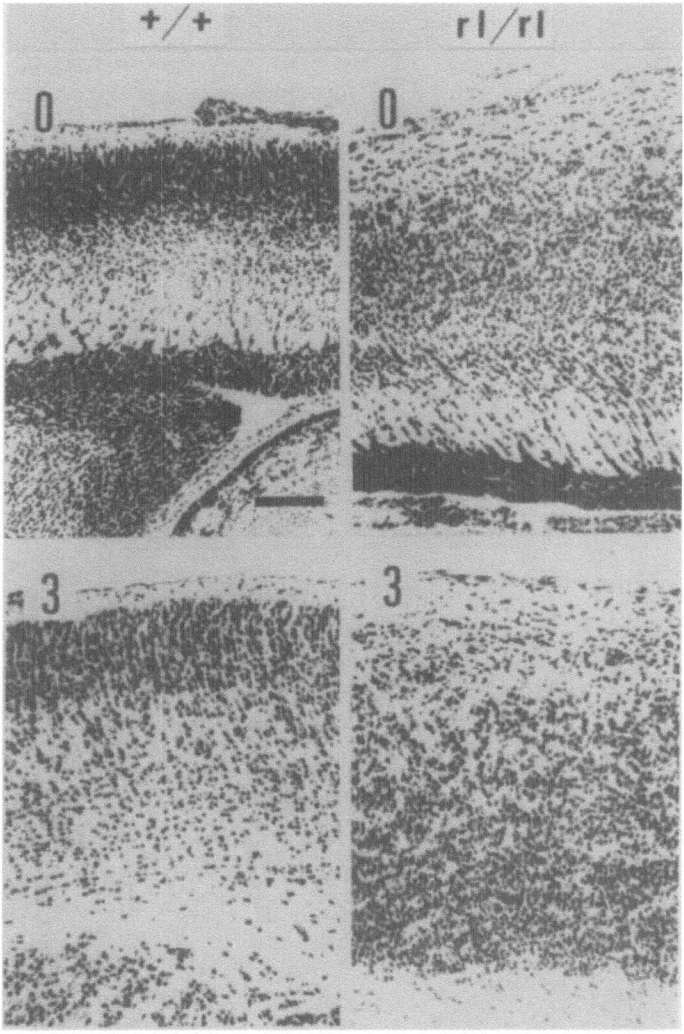

FIG. 2a. Sequential process of formation of the cerebral cortex of control and reeler mice at birth and 3rd postnatal day.

+/+, control; *rl/rl*, reeler; 0, day at birth; 3, 3 days after birth; Upper, sagittal section of the cerebral cortex. Lower, coronal section of the cerebral cortex. Most of the cells have migrated near the surface of the cortex, though not into the
◀ molecular layer, in the control cerebrum. The cells in the

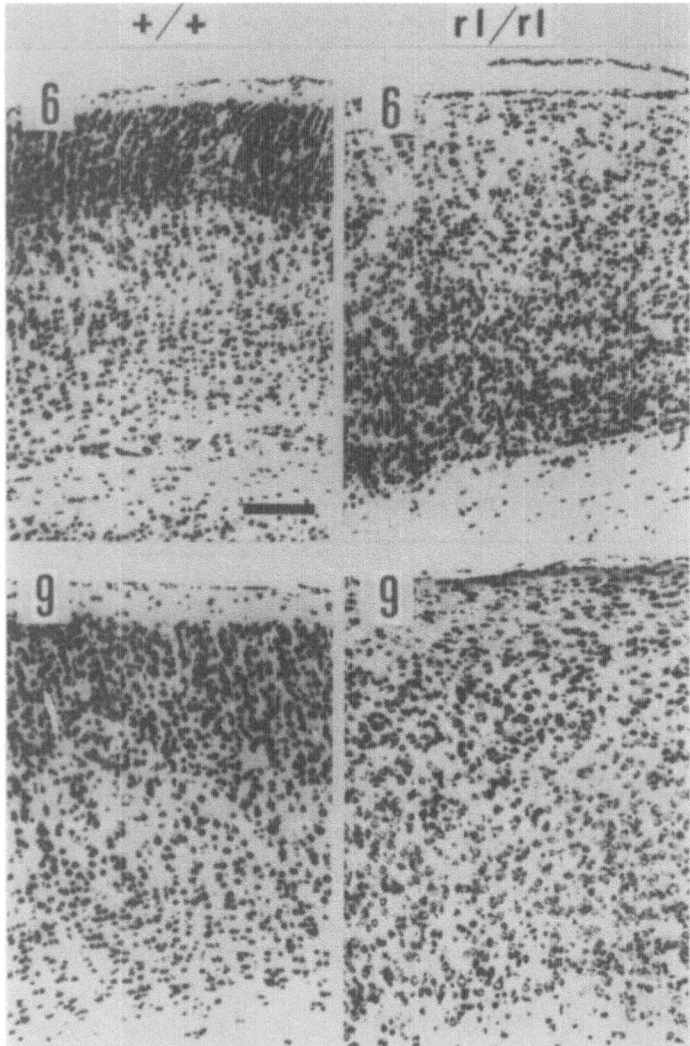

Fig. 2b. Formation of the cerebral cortex of control and reeler mice at 6th and 9th postnatal days. Hematoxylin eosin staining of the coronal sections. The gradient of cell population in the reeler is the opposite of that of the control both at 6th and 9th postnatal days. No molecular layer is observed in the reeler cerebral cortex. (Mikoshiba *et al., J. Neurochem.*, 1980)[12]

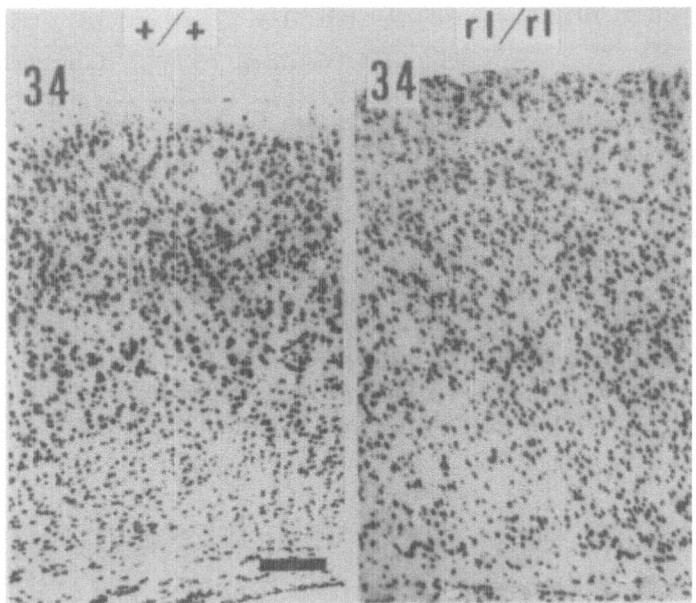

FIG. 2c.   Coronal section of the cerebral cortex from control
and reeler mice at 34 days of age. The cell population is relatively even in both control and reeler. No molecular layer is
observed in the reeler. (Mikoshiba *et al.*, *J. Neurochem.*,
1980)[12]

distributed cell populations with inverted cortical layers (Fig. 2c). Recently,
it was found by fractography that the bundles of fibers through which the
divided cells migrate were abnormal (Mikoshiba *et al.*, in preparation).
Reeler would be a very important model to investigate the mechanism of
cell migration in the mammalian nervous system.

The second question posed on the reeler mutation concerns the mech-
Analyses of several neurochemical markers were performed to show if
and how the inverted laminar cortex affect the genetically determined cell
numbers, number of neuronal terminals, and myelination (Table 1). No
significant difference was found in DNA content, implying that the total
number did not change. No differences were observed in the activity of
2′,3′-cyclic nucleotide 3′-phosphohydrolase (CNPase), a myelin marker
enzyme, and choline acetyltransferase, a marker of cholinergic fibers.
Therefore, the rate of myelination and the number of cholinergic terminals
do not appreciably change in the reeler.

The second question posed on the reeler mutation concerns the mech-

TABLE 1. Neurochemical Changes in the Cerebrum and Cerebellum from Reeler Mutant Mice Compared to Control Mice

	Cerebrum	Cerebellum
DNA content	~	↓
2′,3′-cyclic nucleotide	~	↑
3′-phosphohydrolase		↑
Choline acetyltransferase	~	
Protein profiles	~	↓ histone ↑ P$_{400}$*
Glutamate content	~	↓
2-Deoxyglucose incorporation	~	altered

* P$_{400}$ is a protein characteristic of Purkinje cells in the cerebellum.[30]

anism governing the connection between cells. Immunohistochemical analysis was performed using the antiserum against myelin basic protein, one of the myelin components, to study how the myelinated fiber connec-

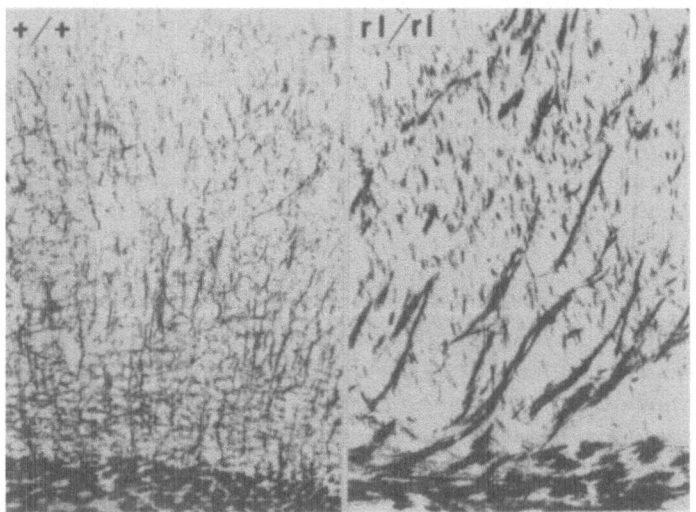

FIG. 3. Higher magnification of the myelinated fibers stained immunohistochemically using antiserum against myelin basic protein. In the control cerebrum, fine mesh of fibers are climbing up vertically to the surface of the cortex. In the reeler, prominent bundles of fibers which climb up in an oblique fashion are observed in addition to the fibers climbing up vertically. Corpus callosum is shown at the bottom of the figure.

tions are altered by inversion of the layers. Myelination is an important marker to judge how the brain functions develop. In the control, myelinated axons went up vertically to the surface of the cerebrum, while in the reeler, there were actually vertical fibers (Fig. 3). In addition, there were densely packed fibers which radiated in oblique fashion to the surface of the cortex and came down gradually (Fig. 3). These bundles of fibers, which are considered to be mainly of thalamic origin, climbed up to the surface and then went down as if they were searching for target cells. Various stages in the appearance of myelinated fibers after birth are shown in Fig. 4. The stage of myelinogenesis did not differ significantly between the control and the reeler. It appears as if the relative position of neurons is not critical to the morphogenetic mechanism that controls the intercellular connections during development.

In order to solve the third question posed on the reeler, that is, how the function of the brain is changed by the genetically altered connections, the 2-deoxyglucose technique was introduced. 2-Deoxyglucose is an analogue

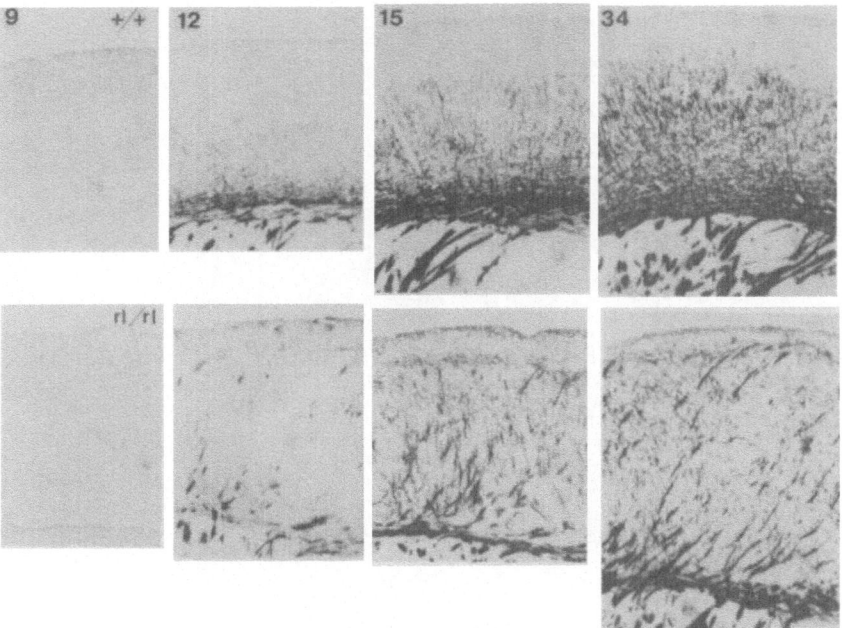

Fig. 4. Immunohistochemical staining of the myelinated fibers in the cerebral cortex at the various postnatal stages. Each number indicates the postnatal days.

of glucose and is incorporated into the cells according to the metabolic requirements of the cells. Since it is not metabolized in the cells, it is possible to visualize the metabolically active part in the autoradiogram by radiolabeling the 2-deoxyglucose. The thalamo-cortical projection terminates at the granular cell layer, the fourth layer, of the cerebral cortex in the control cerebrum.[34,35] The 2-deoxyglucose incorporation was at a maximum at the granular cell layer in the resting stage. The strong incorporation of 2-deoxyglucose was found in the granular cell layer of the reeler revealed by the analysis of the thionine stained sections used for autoradiogram, though the incorporation was relatively more diffuse than in the control (Fig. 5). It is therefore suggested that the contact between the

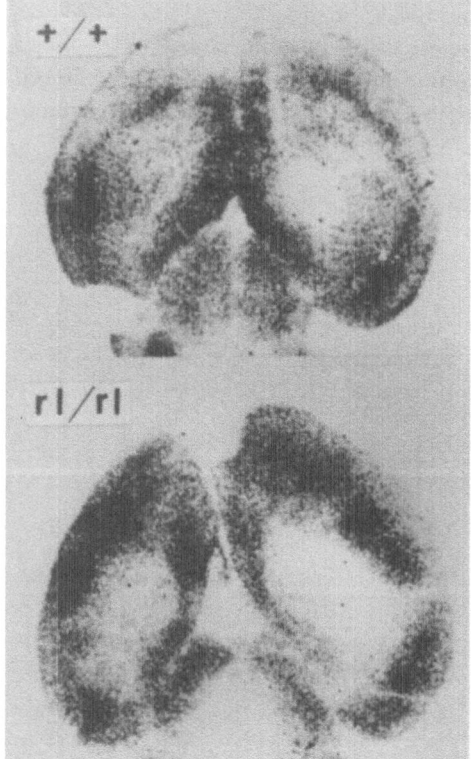

FIG. 5. Metabolic mapping by the 2-deoxyglucose technique of horizontal sections of the cerebrum from control and reeler mice. [14C] 2-Deoxyglucose was injected intraperitoneally, after 45 min, animals were sacrificed and the brains were immediately frozen. (Mikoshiba *et al.*, *J. Neurochem.*, 1980)[12]

thalamocortical projection and the granule cells are working functionally even in an inverted position. It is speculated that during development, most of the cells must have known their specific targets in terms of mutual contact of cells, but not in terms of the direction in which the nerve fibers grow. However, degeneration experiments have revealed another mechanism that delivers terminals to a particular layer of the cortex, independent of the position of target cells.[31]

Development of the cerebellar cortex of the reeler mutant mice

In the adult control cerebellum, Purkinje cells are located between the molecular and granular layers, but in the reeler mutant mice, most of the Purkinje cells exist in the white matter region. Only a few Purkinje cells can be observed at the normal position. Therefore, the cerebellum of reeler mutant mouse is characterized by inversion of Purkinje cells and granule cells.

The postnatal development of the cerebellum is remarkable and the size of the cerebellum increases from the tenth to the twentieth day after birth as is clearly demonstrated in Fig. 6. The reeler cerebellum is reduced in size with little enlargement beyond the sixth day after birth. In the reeler cerebellum, the molecular layer is thinner and the density of the granular layer is much less than the control. It appears that the time schedule of migration of granule cells and the formation of the granular layer in the reeler cerebellum did not differ significantly from that of the control (Fig. 6). The reeler cerebellum contained the same protein profile as the control and had an equal amount of protein.[13] The proteins in reeler increased developmentally in a normal fashion except for histones. Therefore, it is speculated that the reduced size of the reeler cerebellum is not due to an arrest of maturation as far as protein profiles are concerned.

To examine how the function of the reeler cerebellum is altered, the 2-deoxyglucose technique was introduced to the reeler cerebellum. In the control cerebellum, 2-deoxyglucose incorporation is usually highest at the granular layer where mossy fibers, one of the extracerebellar inputs, form synaptic contacts with granule cells. White matter incorporates little 2-deoxyglucose. In the case of reeler mutant mice, 2-deoxyglucose incorporation at the white matter region where most of the Purkinje cells are located was very high compared to that of granular layer (Fig. 7). Even in the cerebellum where Purkinje cells are absent, 2-deoxyglucose incorporation was maximum in the granular layer. Therefore, it appears that mossy fibers rather than climbing fibers, play an important role in 2-deoxyglucose incorporation in the cerebellum. Electrophysiological and electron microscopic studies suggest that heterologous synapses between

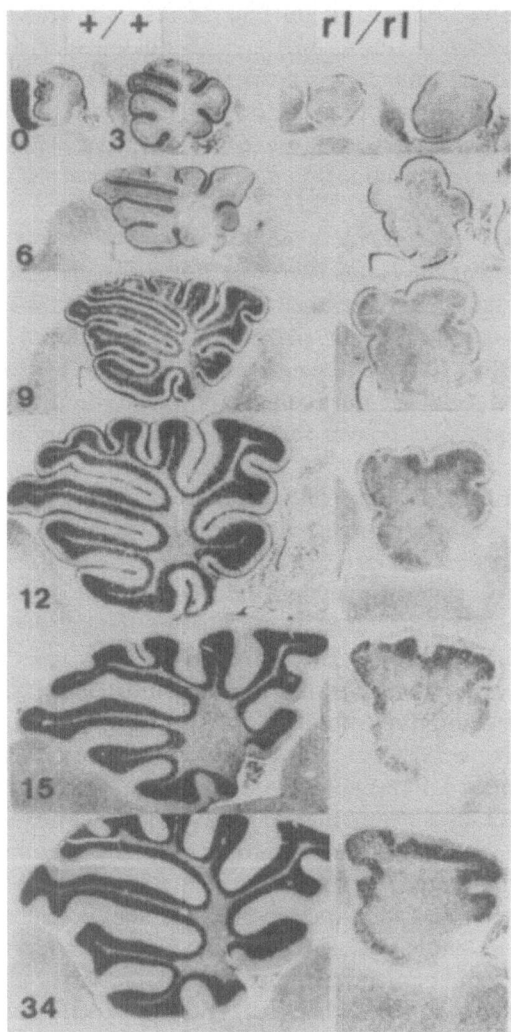

FIG. 6. Sequential postnatal development of the cerebellum. The day of birth is indicated as 0; each number indicates postnatal days. Sections were stained by hematoxylin eosin. The size of the cerebellum did not increase after the 9th postnatal day. In the reeler cerebellum, the formation of lobes is very poor and the cell density of the granular layer is much less than the control. Most of the Purkinje cells are localized in the white matter in the reeler cerebellum. (Mikoshiba *et al.*, *Dev. Biol.*, 1980)[13)]

2 DG   T

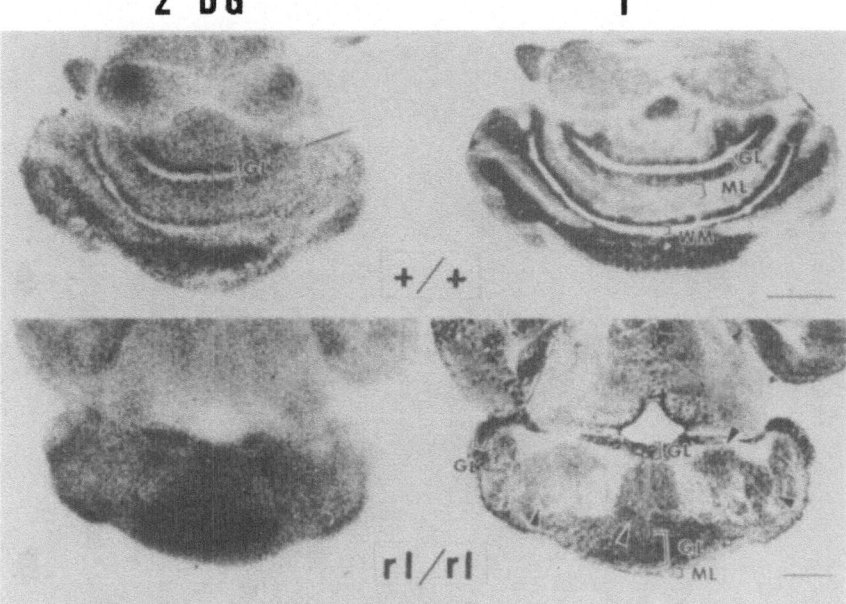

FIG. 7.   Metabolic mapping by the 2-deoxyglucose technique
of the cerebellum from control and reeler mice. 2-DG, auto-
radiograph; T, thionine blue staining of the corresponding
section; ML, molecular layer; GL, granular layer; WM, white
matter. Note that 2-DG is incorporated in the white matter
where most of the Purkinje cells exist (indicated by arrow-
heads), while the maximum incorporation is found in the
granular layer and little incorporation in the white matter in the
control. Bars, 1 mm. (Mikoshiba *et al.*, *Dev. Biol.*, 1980)[13]

mossy fibers and Purkinje cells are found in the reeler cerebellum.[11] Furthermore, the myelinated fiber arborization stained with the antiserum
against myelin basic protein were greatly altered in the reeler (Fig. 8).
Numerous fibers course around the large neurons, probably Purkinje cells,
in various directions in the reeler. From these results, it is suggested that
the cerebellar circuitry was reorganized as a result of the reeler mutation.

A summary of the neurochemical changes in the cerebrum and the cerebellum of the reeler mutant compared to the control is presented in
Table 1. Although no significant differences were observed in the cerebrum,

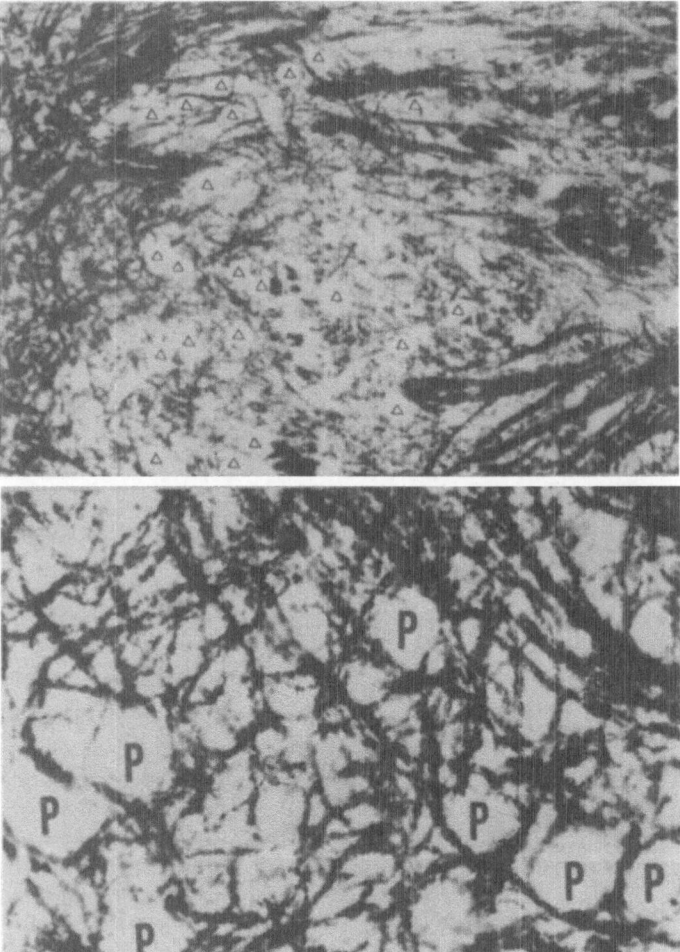

Fɪɢ. 8. Immunohistochemical staining of the myelinated fiber arborization in the reeler mutant cerebellum using the antiserum against myelin basic protein.

Upper figure, lower magnification of the white matter region of reeler mutant cerebellum. Note that the bundles of fibers course in various direction. Numerous white spots of Purkinje cell bodies devoid of staining (indicated by open triangles) are observed.

Lower figure, higher magnification of the white matter region of the reeler cerebellum. Numerous fibers surround the Purkinje cell body (indicated by P).

there were various differences in the cerebellum. There was a remarkable reduction in cell number in the reeler cerebellum as judged from DNA content and histone levels. Since homologous cells are generated simultaneously in reeler and normal mice, the discrepancy between the cerebrum and the cerebellum in the reeler mutant might reflect differences in the schedule of cell division between the two cortices. The cells at the external granular layer usually proliferate to produce granule cells after birth, while most of the neurons in the cerebrum complete their cell division before birth. Inversion of the brain cell layers might damage dividing cells but have little effect on post-mitotic cells.

Another explanation is that the loss of cells *in vivo* in the reeler cerebellum might occur due to the absence of a target, that is, Purkinje cell dendrites, in the molecular layer. If the loss of granule cells in the reeler cerebellum results from the absence of a normal synaptic contact *in vivo*, one would expect that the *in vitro* culture system may promote the growth of granule cells where cellular or fiber contacts are easily performed. Our studies on the growth of explants of the reeler cerebellum in primary culture revealed that neuritic outgrowth, schedule of general development, and quantity of myelin formation were comparable to that of the control except for the lack of laminar structure.[13]

The discrepancy between cerebellar development *in vivo* and *in vitro* in the reeler may be a significant point for elucidating the factors controlling the development of the cerebellum. The presence of the synaptic contact is probably an important factor.

SHIVERER MUTANT MOUSE

The nervous system consists mainly of two types of cells, neurons and glial cells. Glial cells are grouped into astrocytes, oligodendrocytes, and microglia. Oligodendrocytes play the most important role among the glial cells, since oligodendrocytes are involved in myelin formation. The first step of myelinogenesis for the oligodendrocytes is to extend their processes towards the axon, searching for an axon to myelinate. The second step is to recognize and make contact with the axons which are supposed to have been genetically determined. The third stage is to form myelin membranes by molecular assembly of the lipids and proteins, the components of myelin.

Shiverer mutant is an autosomal recessive characterized by hypomyelination of the central nervous system.[15-22] It is genetically different from the other dysmyelinating mutants such as quaking and jimpy mice. The lamella formation of the myelin in shiverer mice is very poor (Figs. 9, 10).

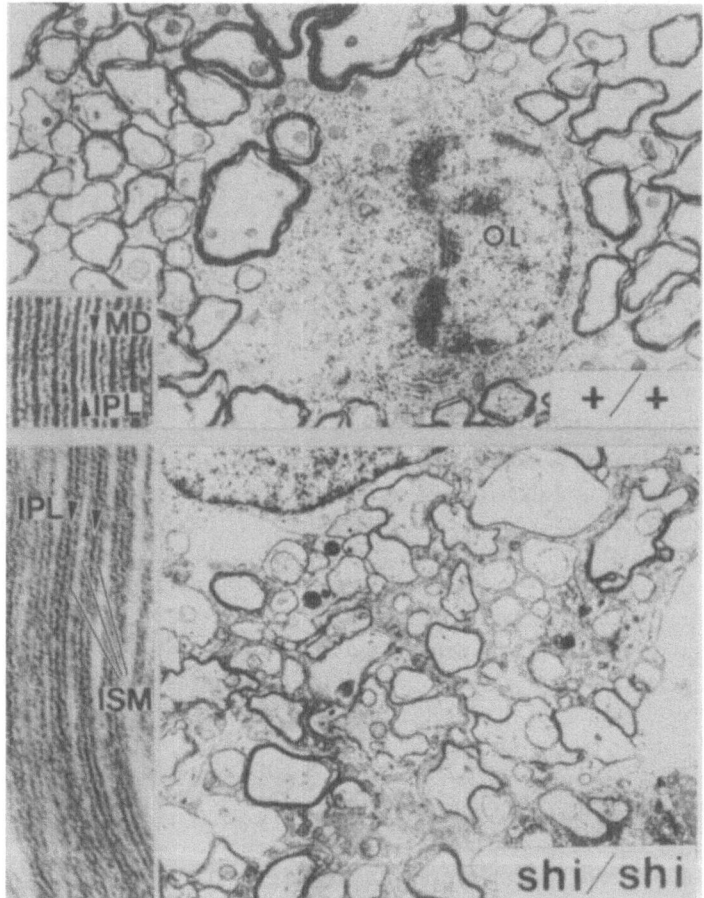

FIG. 9.   Electron micrographs of the tissues from the central nervous system of shiverer and control mice.

Upper figures show the cross section of the spinal cord from control mice. Inset, higher magnification of the control myelin. MD indicates major dense line and IPL indicates intraperiod line.

Lower figures show the cross section of the spinal cord from the shiverer mice. Shiverer is characterized by the poor lamella formation of the myelin sheath. Inset, higher magnification of the shiverer myelin. The myelin is devoid of a major dense line. Note that the inner surface of the unit membrane ▶

The myelin in the shiverer is grouped into two types. In the most common type of myelin formation, the lamella is formed by several layers of cytoplasmic sheets, not spirally wrapping the axon but incompletely enclosed with one turn or less (Fig. 10). The second type of myelin formation, which

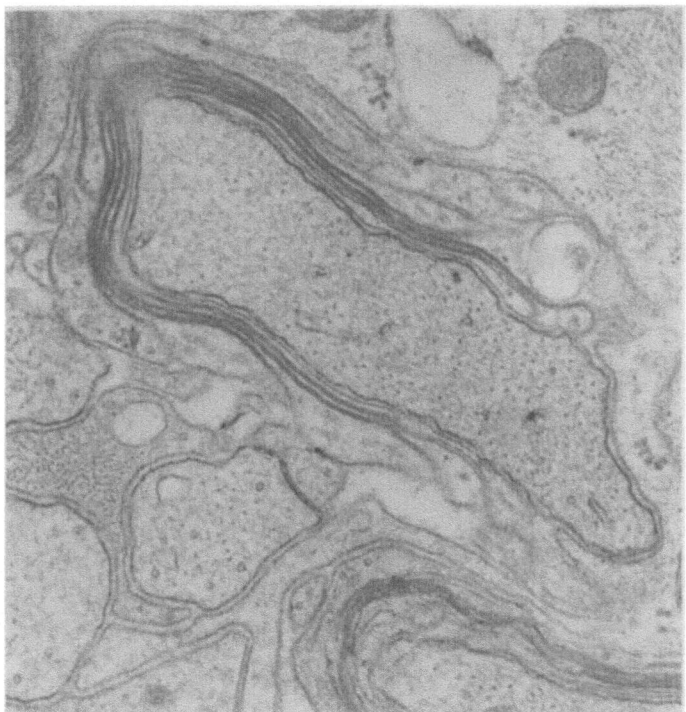

FIG. 10. Electron microscopic image of the myelin in the central nervous system from the shiverer. Section is obtained from the spinal cord. Note that the myelin lamella is formed by several layers of the cytoplasmic sheets of oligodendrocyte, not spirally wrapping but incompletely enclosing the axon with one turn or less.

◀ (indicated as ISM) of the oligodendrocyte processes does not fuse to form a single major dense line. Intraperiod line (IPL) is formed as a result of the close apposition of the outer leaflets of the unit membrane.

is rarely observed, is an incomplete type of myelin sheath spirally but loosely wrapping the axon. By careful analysis of the figures of the first major type of myelination, it was found that the cytoplasmic processes were frequently split into several layers and enclosed the axons by piling up in layers.[23,33] Some cytoplasmic sheets in the piles are often split into several layers to enclose the same axon.[23] It is likely that the recognition step of axons by the processes of oligodendrocytes are normal but the subsequent step of wrapping the axons is affected by the mutation.

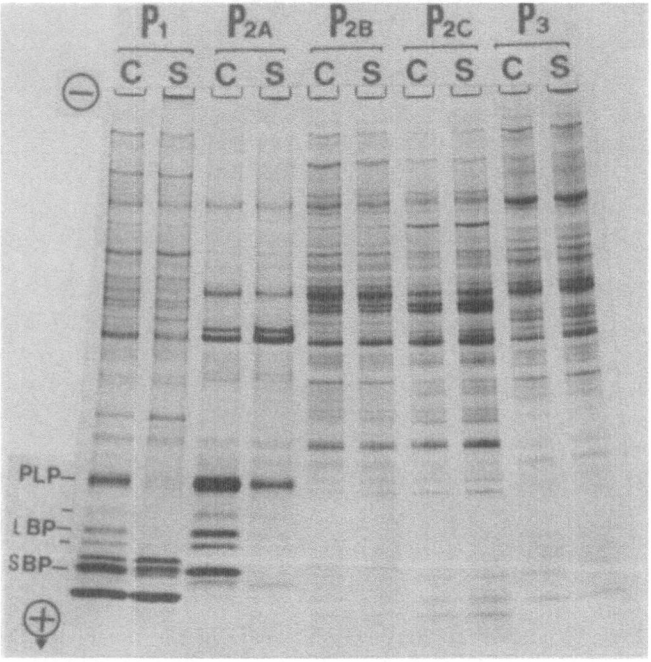

FIG. 11. Protein profiles of the fractions after subcellular fractionation of the homogenates of the brains from shiverer and control mice. C, fractions from the controls; S, fractions from the shiverer; P_1, crude nuclear fraction; P_{2A}, myelin fraction; P_{2B}, synaptosomal fraction; P_{2C}, mitochondrial fraction; P_3, microsomal fraction. The positions where proteolipid protein, large basic protein, and small basic protein migrate are indicated by PLP, LBP, and SBP. LBP and SBP and other two bands are nearly lost in the shiverer. Histone bands overlap to the SBP in P_1 fraction, recognized both in C and S. (Mikoshiba *et al.*, *J. Neurochem.*, 1980)[19]

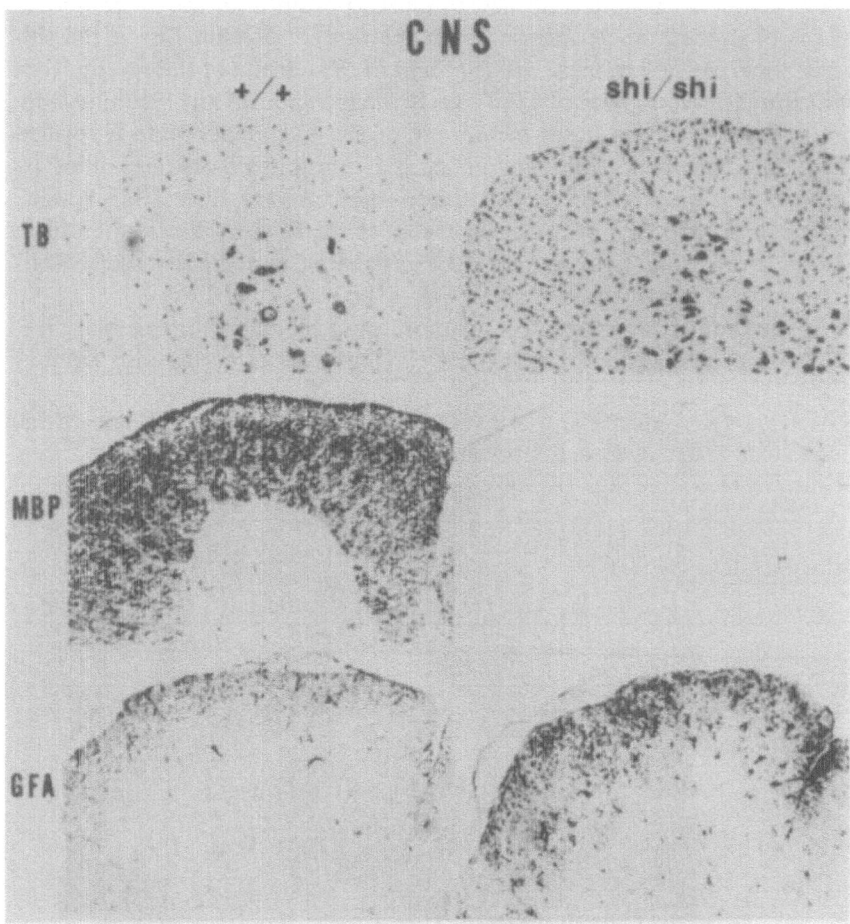

FIG. 12. Toluidine blue (TB) staining and immunohistochemical staining by the antibodies against myelin basic
protein (MBP) and glial fibrillary acidic (GFA) protein of the
cross section of the spinal cord from control and shiverer
mice. Note that numerous cells are found in the white matter
of the shiverer. Immunohistochemical staining shows a positive reaction with the white matter, while no reaction is found
in the shiverer. Numerous fibers are observed in the white
matter of the spinal cord from the shiverer. Most of the increased cells are negative to the S-100 staining, namely no
significant differences were observed between the control and
the shiverer.

The major dense line is formed by close apposition of the inner surface of the oligodendrocytic membrane in the normal myelin. But in the shiverer mice, the major dense line is absent in the lamella of the myelin from the central nervous system[18-20,23]. Only the intraperiod line resulting from the apposition of the outer membrane of the oligodendrocyte is formed (Fig. 9). Subcellular fractionation of the brains from shiverer mice by sucrose density gradient centrifugation showed very interesting results. Myelin basis proteins were clearly absent in the P_{2A}, myelin fraction from the shiverer. Proteolipid protein (PLP) appeared to decrease, while Wolfgram protein relatively increased (Fig. 11).

By the immunohistochemical reaction using the antiserum against myelin basic protein, the absence of myelin basic protein was also demon-

TABLE 2. The Activities of 2′,3′-Cyclic Nucleotide 3′-Phosphohydrolase of the Subcellular Fractions from Control and Shiverer Mutant Mice

		(CNS)		(PNS)
		CNPase		CNPase
Subcellular fractions		μ mol/min/ mg protein	μ mol/min/ fraction	μ mol/min/ mg protein
P_1	+/?	2.37	40.3	0.32
	shi/shi	0.67 (28%)	10.3 (26%)	0.31
P_2	+/?	2.73	139.8	1.83
	shi/shi	2.58 (94%)	111.2 (80%)	1.91
P_{2A}	+/?	16.44	59.2	2.12
	shi/shi	14.84 (90%)	8.5 (14%)	2.30
P_{2B}	+/?	0.70	5.5	1.69
	shi/shi	3.17 (452%)	27.4 (498%)	2.06
P_{2C}	+/?	0.14	1.4	0.29
	shi/shi	0.21 (150%)	2.0 (142%)	0.23
P_3	+/?	1.10	26.9	1.69
	shi/shi	3.72 (338%)	109.6 (407%)	1.62
P_{3X}	+/?			3.67
	shi/shi			3.58
P_{3A}	+/?			2.40
	shi/shi			2.27
S	+/?	0.04	0.3	
	shi/shi	0.04 (100%)	0.3 (100%)	
Purified	+/?	27.76		3.08
myelin	*shi/shi*	29.15		2.92

Although no significant difference was found in the activity of the control and shiverer in the PNS, prominent differences are found in the CNS. (Mikoshiba *et al.*, *J. Neurochem.*, 1980)[19]

strated[24] (Fig. 12). Since myelin basic protein was not detected in any fraction of the brain from the shiverer mice and the immunohistochemical reaction was negative throughout the development, it is suggested that the *de novo* synthesis of basic protein is inhibited by the shiverer mutation (Fig. 13).

2′,3′-Cyclic nucleotide 3′-phosphohydrolase, CNPase, is considered to be a myelin marker enzyme and the activity correlated well with the level

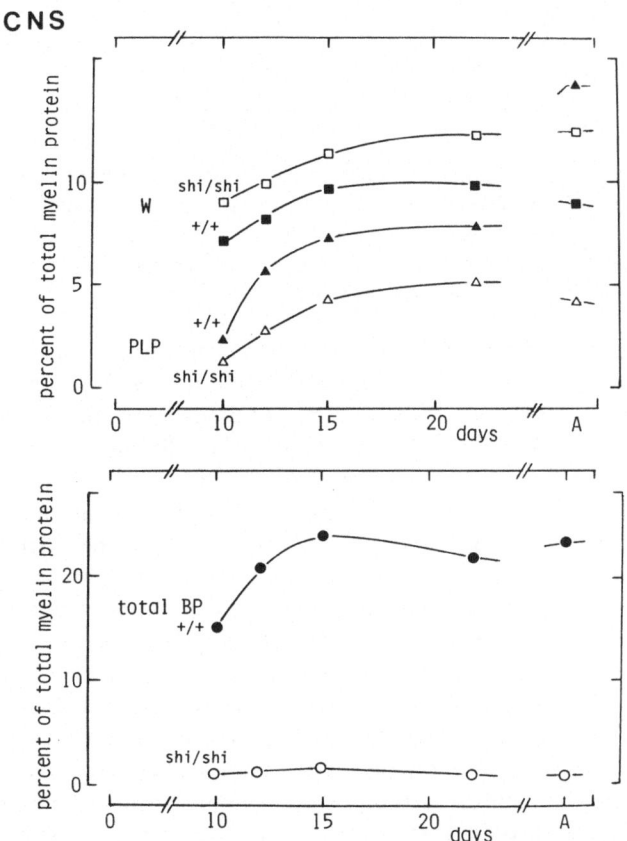

FIG. 13a. Changes in protein compositions of the CNS from shiverer and control mice during postnatal development. The distribution of myelin proteins were quantified by densitometry. The major myelin proteins were plotted as percentages of the total proteins. W, Wolfgram protein; PLP, proteolipid protein; total BP, total amount of MBP (myelin basic protein).

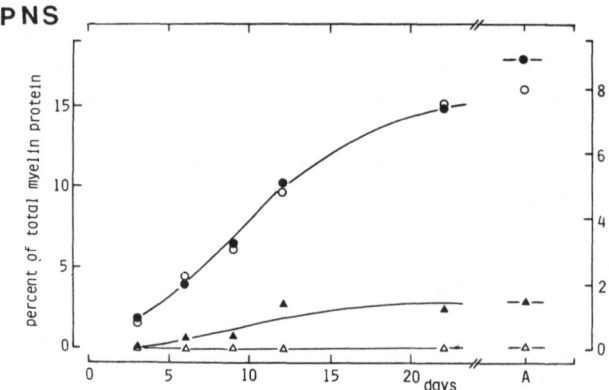

FIG. 13b.　Changes in protein composition of sciatic nerve fibers from shiverer and control mice during postnatal development. P0 protein: control (●), shiverer (○); MBP: control (▲), shiverer (△). The myelin proteins were plotted as percentages of the total proteins.

of myelination.[25-27] CNPase activity of the central nervous system from other dysmyelinating mutants, jimpy and quaking, were much lower than the control.[28,29] The activity of CNPase of tissue from the shiverer was as high as that of the control despite the poor lamella formation in the shiverer (Table 2). The CNPase activity of purified myelin was nearly the same as that from the control. It is clear that some of the CNPase molecules are located within the myelin sheath in the shiverer. The enrichment of CNPase activity in microsomal P_3 and synaptosomal P_{2B} fractions revealed that CNPase molecules also exist in (or on) the membrane such as premyelin or rough surfaced endoplasmic reticulum.

It has been demonstrated that there is a clear difference in the process of forming the myelin sheath between the Schwann cell and oligodendrocyte. The Schwann cell body goes around the axons to wrap them, and the myelin components in the peripheral nervous system are very different from that in the central nervous system. Wolfgram protein, PLP, and small and large myelin basic proteins are the main components in the central nervous system, while in the peripheral nervous system, P0 protein is the major component and P1, P2 and Pr are included as minor components (Fig. 14). It is clear that the process of molecular assembly to form myelin lamella is quite different between the central and the peripheral nervous system.

By the subcellular fractionation technique, it was found that P1, PM,

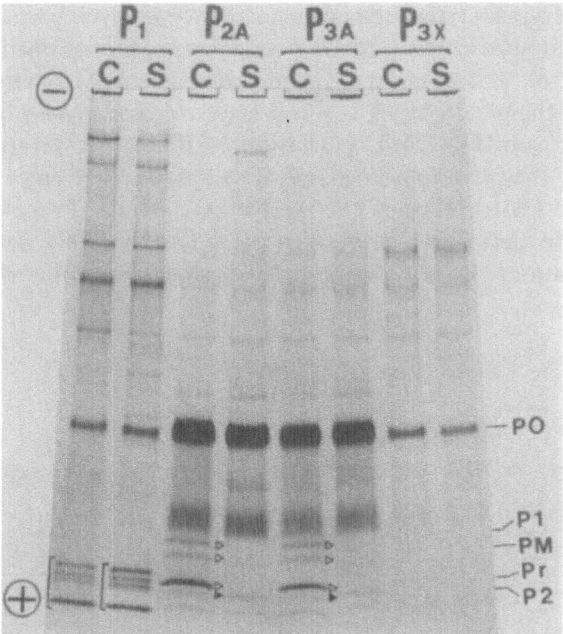

Fig. 14. Protein profiles of the subcellular fractions of sciatic nerve from shiverer (s) and control (c) mice. P_1, crude nuclear fraction; P_{2A}, P_{3A}, fractions recovered at the interface between 0.32–0.8 M sucrose after centrifugation of the P_2 and P_3 fractions on the discontinuous sucrose gradient (0.32–0.8–1.2 M sucrose); P_{3X}, fraction floating on 0.32 M sucrose after centrifugation.

Pl, PM, Pr proteins which are myelin basic proteins (MBPs) common to CNS were missing in the shiverer but P2 protein which is a specific MBP of PNS existed.

and Pr proteins were missing in both P_{2A} and P_{3A} fractions in the shiverer where myelin from the peripheral nervous system was recovered (Fig. 14). No extra bands were observed in the P_{2A} and P_{3A} of the shiverer mice. The CNPase activity was high in the fractions which contained a high percentage of myelin proteins, except P_{3X} fraction which would be a precursor of the myelin components.[20] Pr protein corresponds to the small basic protein of the central nervous system, while Pl protein corresponds to the large myelin basic protein.[21] Among the myelin components from the peripheral nervous system of the shiverer, both Pl and Pr proteins, common to central myelin, were missing but P0 and P2 proteins specific to the peripheral nervous system remained at the same level as in the controls (Fig. 14). The myelin basic protein was absent throughout the development (Fig. 13b). The absence of myelin basic protein was also confirmed immunohistochemically using the antiserum against myelin basic protein of central origin (Fig. 15). Localization of P2 in the shiverer mice was confirmed by the immunohistochemical technique.[32] Even in the absence of P1, Pr, and PM proteins in the myelin, the lamella formation proceeds equally in the peripheral nervous system. In the CNS, however, poor myelin formation with the absence of the major dense line was observed

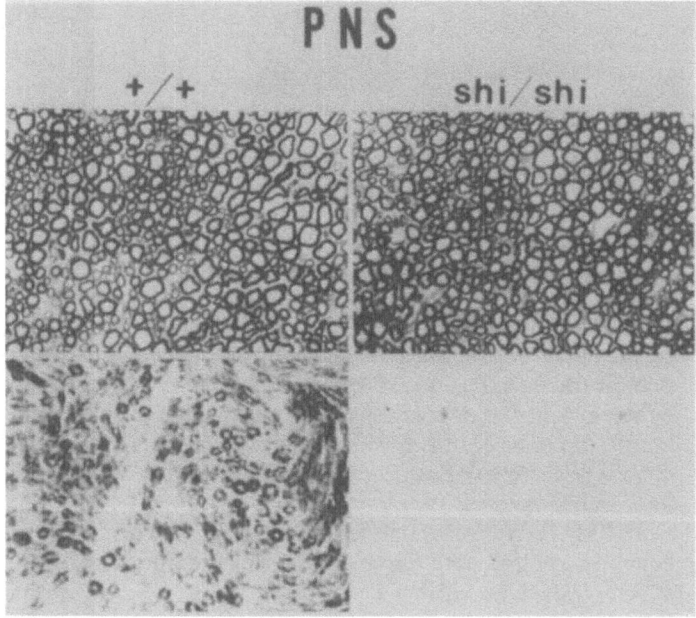

FIG. 15. Immunohistochemical reaction on the sections of the sciatic nerve fibers from shiverer and control mice with the antiserum against myelin basic protein (MBP).

Upper: Osmium staining of the 1 μm thick sections from the control and the shiverer.

Lower: Reaction with the antiserum against MBP on the sections of the sciatic nerve fibers.

Note that the lamellar formation of the sciatic nerve from the shiverer does not show any significant difference from the control and that no reaction was observed on the myelin from the shiverer with the antiserum against MBP.

accompanying the absence of small and large myelin basic proteins. This difference would be a significant clue to elucidate the mechanism of myelin formation which is the result of molecular assembly of the myelin components.

REFERENCES

1. Jacobson, M.: Developmental Neurobiology (2nd edition). Plenum Press, New York, 1978.
2. Changeux, J.-P. and Mikoshiba, K.: Genetic and 'epigenetic' factors regulating synapse formation in vertebrate cerebellum and neuromuscular junction. In: Maturation of the Nervous System: Progress in Brain Research, Vol. 48 (ed. M. A. Corner, R. E. Baker, N. E. van de Pol, D. F. Swaab and H. B. M. Uylings), pp. 43–66. Elsevier Scientific, Amsterdam, 1978.
3. Sidman, R. L., Green, M. C. and Appel, S. H.: Catalog of the Neurological Mutants of the Mouse. Harvard University Press, Cambridge, Massachusetts, 1965.
4. Sidman, R. L.: Cell interactions in developing mammalian central nervous system. In: Cell Interactions. Proceedings of the Third Lepetit Colloquium (ed. L. G. Silvestri), pp. 1–13. North-Holland, Amsterdam, 1972.
5. Falconer, D. S.: Two new mutants, "trembler" and "reeler," with neurological actions in the house mouse (*Mus musculus L.*). *J. Genet.*, **50**: 192–201, 1951.
6. Hamburgh, M.: Observations on the neuropathology of "reeler," a neurobiological mutation in mice. *Experientia*, **16**: 460–461, 1960.
7. Hamburgh, M.: Analysis of the postnatal developmental effects of "reeler," a neurological mutation in mice: A study in developmental genetics. *Develop. Biol.*, **8**: 165–185, 1963.
8. Caviness, V. S., Jr.: Patterns of cell and fiber distribution in the neocortex of the reeler mutant mouse. *J. Comp. Neurol.*, **170**: 435–448, 1976.
9. Caviness, V. S., Jr. and Rakic, P.: Mechanisms of cortical development: A view from mutations in mice. *Ann. Rev. Neurosci.*, **1**: 297–326, 1978.
10. Caviness, V. S., Jr. and Sidman, R. L.: Time of origin of corresponding cell classes in the cerebral cortex of normal and reeler mutant mice: An autoradiographic analysis. *J. Comp. Neurol.*, **148**: 141–152, 1973.
11. Mariani, J., Crepel, F., Mikoshiba, K., Changeux, J.-P. and Sotelo, C.: Anatomical, physiological and biochemical studies of the cerebellum from reeler mutant mouse. Philosophical Transactions of the Royal Society. B. *Biological Sciences*, **281**: 1–28, 1977.
12. Mikoshiba, K., Kohsaka, S., Takamatsu, K., Aoki, E. and Tsukada, Y.: Morphological and biochemical studies on the cerebral cortex from reeler mutant mice: Development of cortical layers and metabolic mapping by the deoxyglucose method. *J. Neurochem.*, **34**: 835–844, 1980.
13. Mikoshiba, K., Nagaike, K., Kohsaka, S., Takamatsu, K., Aoki, E. and

220 K. Mikoshiba *et al.*

Tsukada, Y.: Developmental studies on the cerebellum from reeler mutant mouse *in vivo* and *in vitro*. *Dev. Biol.*, **79**: 64–80, 1980.

14. Stewart, R. M., Richman, D. P. and Caviness, V. S., Jr.: Lissencephaly and pachygyria: An architectonic and topographical analysis. *Acta Neuropath.* (Berl.), **31**: 1–12, 1975.

15. Biddle, R., March, E. and Miller, J. R. *Mouse News Lett.*, **48**: 24, 1973.

16. Bird, T. D., Farrell, D. F. and Sumi, S. M.: Brain lipid composition of the shiverer mouse: Genetic defect in myelin development. *J. Neurochem.*, **31**: 387–391, 1978.

17. Chernoff, G., March, E. and Miller, J. R. *Mouse News Lett.*, **51**: 12, 1974.

18. Privat, A., Jacque, C., Bourre, J. M., Dupouey, P. and Baumann, N.: Absence of the major dense line in myelin of the mutant mouse "shiverer". *Neuroscience Letters*, **12**: 107–112, 1979.

19. Mikoshiba, K., Nagaike, K. and Tsukada, Y.: Subcellular distribution and developmental change of 2′,3′-cyclic nucleotide 3′-phosphohydrolase in the central nervous system of the myelin-deficient shiverer mutant mice. *J. Neurochem.*, **35**: 465–470, 1980.

20. Mikoshiba, K., Nagaike, K., Takamatsu, K. and Tsukada, Y.: Developmental change of 2′,3′-cyclic nucleotide 3′-phosphohydrolase activity in the nervous system of the *shiverer* mutant mice *in vivo* and *in vitro*. In: Neurological Mutants Affecting Myelination: Research Tool in Neurobiology (ed. N. Baumann), pp. 349–354. Elsevier Scientific, Amsterdam, 1980.

21. Mikoshiba, K., Kohsaka, S., Takamatsu, K. and Tsukada, Y.: Neurochemical and morphological studies on the myelin of peripheral nervous system from Shiverer mutant mice: Absence of basic proteins common to central nervous system. *Brain Res.*, **204**: 455–460, 1981.

22. Mikoshiba, K., Aoki, E. and Tsukada, Y.: 2′,3′-Cyclic nucleotide 3′-phosphohydrolase activity in the central nervous system of a myelin deficient mutant (shiverer). *Brain Res.*, **192**: 195–201, 1980.

23. Inoue, Y., Nakamura, R., Mikoshiba, K. and Tsukada, Y.: Fine structure of the central myelin sheath in a myelin deficient mutant "shiverer", with special reference to the pattern of myelin formation by oligodendroglia. *Brain Res.*, **219**: 85–94, 1981.

24. Dupouey, P., Jacque, C., Bourre, J. M., Cesselin, J., Privat, A. and Baumann, N.: Immunochemical studies of myelin basic protein in shiverer mouse devoid of major dense line of myelin. *Neuroscience Letters*, **12**: 113–118, 1979.

25. Kurihara, T. and Tsukada, Y.: The regional and subcellular distribution of 2′,3′-cyclic nucleotide 3′-phosphohydrolase in the central nerovus system. *J. Neurochem.*, **14**: 1167–1174, 1967.

26. Kurihara, T. and Tsukada, Y.: 2′,3′-Cyclic nucleotide 3′-phosphohydrolase in the developing chick brain and spinal cord. *J. Neurochem.*, **15**: 827–832, 1968.

27. Tsukada, Y., Nomura, M., Nagai, K., Kohsaka, S., Kawahata, H. and Ito, M.: Neurochemical correlates of learning ability. In: Behavioral Neurochemistry, pp. 63–84. Spectrum Publications, Inc., Jamaica, New York, 1977.

28. Kurihara, T., Nussbaum, J. L. and Mandel, P.: 2′,3′-Cyclic nucleotide 3′-phosphohydrolase in brains of mutant mice with deficient myelination. *J. Neurochem.*, **17**: 993–997, 1970.

29. Mikoshiba, K., Nagaike, K., Aoki, E. and Tsukada, Y.: Biochemical and immunohistochemical studies on dysmyelination of quaking mutant mice *in vivo* and *in vitro*. *Brain Res.*, **117**: 287–299, 1979.

30. Mikoshiba, K., Huchet, M. and Changeux, J.-P.: Biochemical and immunological studies on the P_{400} protein, a protein characteristic of the Purkinje cell from mouse and rat cerebellum. *Devel. Neurosci.*, **2**: 254–275, 1979.

31. Caviness, V. S., Jr.: Reeler mutant mouse: A genetic experiment in developing mammalian cortex. In: Approaches to the Cell Biology of Neurons (ed. W. M. Cowan and J. A. Ferrendelli): Society for Neuroscience Symposia, Vol. II, pp. 27–46.

32. Mikoshiba, K., Takamatsu, K. and Tsukada, Y.: Peripheral nervous system of the shiverer mutant mice: Developmental change of myelin components and immunohistochemical demonstration of the absence of MBP and presence of P2 protein. (Submitted)

33. Inoue, Y., Nakamura, R., Mikoshiba, K. and Tsukada, Y.: The formation patterns of central myelin sheaths in the myelin deficient mutant shiverer mouse. *Okajima Folia Anatomica Japonica*, **58**: 613–626, 1982.

34. Wise, S. P. and Jones, E. G.: Developmental studies of thalamocortical and commissural connections in the rat somatic sensory cortex. *J. Comp. Neur.*, **178**: 187–208, 1978.

35. Caviness, V. S., Jr. and Frost, D. O.: Tangential organization of thalamic projections to the neocortex in the mouse. *J. Comp. Neur.*, **194**: 335–367, 1980.

Mutations Affecting Myelination: Dynamics of CNS Myelin Markers in Normal and Dysmyelinating Mutant Mice

B. Zalc, C. Jacque, and N. Baumann

Development of CNS occurs by successive or concomitant differentiation steps, among which myelinogenesis is one of the latest to take place and is the most impressive from the point of view of the amount of newly synthesized material it produces.

Myelin was first described in the nineteenth century by Virchow, and its functions as an electrical insulator had already been postulated in 1878 by Ranvier. This French anatomist was also the first to describe the interruptions (known as "nodes of Ranvier") of the myelin sheaths, which occur at regular intervals along the myelinated axons. These gaps are of crucial importance as they are the only areas of the axonal membrane exposed to depolarization during the conduction of the nerve impulse. This gives rise to what is known as the saltatory conduction. This type of electrical conduction, when compared to the continuous type of propagation of the nerve impulse in the unmyelinated axons, presents two incomparable advantages: it saves energy and speeds the conduction. As stated by Morell and Norton in their excellent review: "If human spinal cord contained only bare nerve fibers, it would have to be several yards in diameter to maintain its conduction velocities."[1]

We will first briefly summarize the structural and metabolic studies obtained by both morphologists and biochemists. This basic knowledge is important for an understanding of how the analytical studies have led to a more dynamical approach to myelinogenesis, involving cellular differentiation processes and neuronal-glial symbiotic interactions. The contribution of neurological murine mutations affecting myelinogenesis will then be analyzed.

STRUCTURE OF NORMAL MYELIN

In this section, in agreement with the editor's wish, we have kept re-

Laboratoire de Neurochimie, INSERM U-134, Hôpital de la Salpêtrière, Paris, France

ferences to a minimum. A detailed bibliography may be found in several recently published reviews.[2-7]

Morphology

Early work using polarized light and X-ray diffraction has already shown the highly ordered structure of myelin. Such studies have demonstrated that, within the sheath, lipid molecules are arranged with their long axes oriented radially and that there is a repeating unit in the radial direction. Electron microscopic studies have shown that the myelin sheath is constituted by the juxtaposition of lamellae disposed in spirals, thus explaining the repetition of a faint line (the intraperiod line), alternating with a darker line (the major dense line). Myelin has been shown to be synthesized by glial cells (Schwann cells in the PNS and oligodendrocytes in the CNS). Myelination thus consists of the wrapping of cytoplasmic processes from these glial cells around the axons. The intraperiod lines correspond to the apposition of the two outer faces of the plasma membrane of this process, while the major dense line is formed by the fusion of the inner or cytoplasmic surface of the same process. Both dark and light lines are thus proteinaceous. These lines are separated by lipid bilayers.

Each internodal segment of myelin is formed by a single cell. In the PNS, one Schwann cell myelinates only one internode of one axon; in the CNS, one oligodendrocyte sends processes to several axons. The formation of the major dense line results in the nearly total exclusion of cytoplasm from the myelin sheath. Indeed, cytoplasm is confined to the periphery (marginal cytoplasmic belt) and to longitudinal and circumferential incisures communicating with the marginal cytoplasmic belt. These cytoplasmic incisures are probably of great importance as they are the only link of the extremity of the myelin lamellae with the cytoplasmic machinery of the mother cell.

Biochemistry

When one compares the molecular composition of myelin to that of any other type of cellular or subcellular membrane, the most striking characteristic is its richness in lipids. This property has been taken advantage of to purify myelin by density gradient centrifugation. Indeed, analysis of isolated myelin shows that lipids account for 70% (for the CNS) of the dry weight. These lipids can be divided in three groups: cholesterol, phospholipids, and glycolipids.

Cholesterol is found in a slightly higher concentration in myelin than in other membranes. Enzymes in the synthetic pathway for cholesterol increase in activity during myelination and then decrease. Although cholesterol esters are mainly present at an early stage of myelination, the activity of cholesterol-esterhydrolase remains at a maximal level.

The concentration of phospholipids in myelin is less than in other membranes. While ubiquitous phospholipids such as phosphatidyl-choline and phosphatidyl-ethanolamine are also found in myelin, triphosphoinositide and ethanolamine plasmalogen are enriched in myelin. Except for the enzymes which synthesize phosphatidyl-choline and phosphatidyl-inositol, developmental changes in many of the enzymes which metabolize individual phospholipids and their cellular and subcellular locations in the brain have not been extensively documented. Several enzymes rise and fall in parallel with the rate of myelination, among them glycerol-phosphate dehydrogenase and acyltransferases.

For sphingomyelin, there seem to be different pools according to molecular species, which might be related to their cellular and subcellular location, oligodendroglial cells and myelin being richer in long chain fatty acid sphingomyelins.

Typical of myelin is its high content of glycolipids. Among them galactosylceramide is the most characteristic. It represents 20% of the total myelin lipids. Galactosylceramide is highly immunogenic, and it has been identified as the so-called "lipid hapten of the nervous system."[8] Galactosylceramide is so specific for myelin and oligodendrocyte that it is now commonly used by tissue and cell cultivators as an oligodendroglial marker.[9,10] The fatty acid composition of this glycosphingolipid is also characteristic. Indeed, its ceramide moiety is extremely rich in very long chain, saturated or monoenoic fatty acids, part of them being α-hydroxylated. No other animal membrane system has such high concentrations of this glycolipid. It must be presumed that it is concentrated in myelin for some structural purpose that may be related in part to the very long acyl chains, as their presence in the myelin sheath confers to this membrane an increasing stability. The synthesizing enzyme of galactosylceramide (the UDP-galactose: ceramide galactosyl transferase) is located in white matter microsomes. The enzyme is most active at the time of myelination, whether it leads to the synthesis of non-hydroxy or hydroxy fatty acids.

Although more ubiquitous than galactosylceramide, its sulfate ester, 3-sulfogalactosylceramide, is highly enriched in myelin and accumulates rapidly as myelination begins. The PAPS cerebroside sulfotransferase, the enzyme responsible for its synthesis, also increases rapidly in activity while myelin is formed. Two gangliosides (sialylated glycolipids) have been described in the myelin: G_{M1} (II^3-NeuAc-Ggose4-Cer) which is the major ganglioside in brain mature myelin and NeuAc Gal-Cer. The latter, which increases with aging in human and rodent, is found in large amount in bird's myelin. Other glycolipids which have been found associated with myelin are monogalactosyldiglyceride and sulfogalactosylglycerolipid.

The protein composition of myelin is also very characteristic. Indeed, as already stated, myelin differs from other plasma membranes by a low content of protein and also by the relative simplicity of its protein profile. Myelin basic protein (MBP) and proteolipid protein account for nearly 70% of the protein content of the CNS myelin.

In most mammals, MBP separates as one major band in electrophoresis. In rat and mouse, two basic proteins can be differentiated, a larger (LBP) and a smaller (SBP) component. Barbarese *et al.*[11] have demonstrated that these two bands of basic protein could again be separated in two: the pre-small and the pre-large. Thus, in the mouse brain, there are four structurally related MBP whose molecular weights are: Pre LBP 21.5 K, LBP 18.5 K, Pre SBP 17 K, and SBP 14 K. Injection of MBP with complete Freund adjuvant to rabbits or guinea pigs results in a lymphocyte-mediated disease known as experimental allergic encephalomyelitis. Immunohistochemical studies have shown that MBP is present in the oligodendrocyte prior to the deposition of myelin.

After myelin formation has begun, MBP can no longer be detected by these techniques in the oligodendroglial cells. MBP has been shown to be localized at the major dense line. We will study below the contribution of the shiverer mutant, lacking immunoreactive MBP, in this finding.

The proteolipid protein (PLP) is the major protein of CNS myelin. Its molecular weight is 25 K. Acidic lipids such as sulfogalactosylceramide, phosphatidyl-serine, and polyphosphoinositides are bound to the apoprotein through ionic linkage, while fatty acids are bound by ester linkage to the polypeptide chain. Due to this lipidic environment, PLP is soluble in chloroform-methanol. PLP has been thought to be localized at the intraperiod line.

In the 20 K region, on polyacrylamide gels, there exists between the LBP and the PLP, a minor component known as the DM20. In some conditions, this intermediate protein appears as a doublet; the smaller component has been identified as a basic protein, while the larger is suspected to be related to the PLP.

In the 50 K region, two acidic proteins exist which were first described by Wolfgram. Several arguments tend to identify these proteins with the 2',3'-cyclic nucleotide 3'-phosphohydrolase. But this point is still controversial. The other myelin high-molecular-weight proteins have not been yet subject to extensive studies. They are thought to be related with the different enzymatic activities formed in purified myelin.

Among the high-molecular-weight proteins, the 100 K myelin associated glycoprotein (MAG) seems of special interest. Indeed, it has been localized immunohistologically in the oligodendrocytes before the start of myelination. It could not be detected in compact myelin, but it has been localized in periaxonal regions of oligodendroglia. This suggests that the MAG might play a role as a recognition signal between the axon and the oligodendroglial process.

The high stability of myelin, which has already been stated, is emphasized by turnover studies. Indeed, the half-life of myelin lipids has been calculated to be in the order of magnitude of weeks, while the estimated half-life of major myelin proteins is several months.

Sequences of myelination

Myelinogenesis occurs at different developmental stages in different species. Thus, the guinea pig has nearly achieved its myelination at birth, while in humans myelination of the CNS begins at birth and some areas are still not myelinated at the end of the second year. In the rat and mouse, myelination starts at around 5 days post-natal and ends at 30–60 days. Furthermore, in a given species, there is a dyschrony between the various parts of the CNS. Thus, the peak of activity of myelin formation, usually reported at 18–21 days in the mouse, is only an average, as regional studies have shown that there is a caudo-rostral timing of myelination (see below).

At the cellular level, it has been shown that myelination starts by the migration of oligodendroglial cells in the bundles of axons to be myelinated. These cells then have a typical in-row disposition, known as the intrafascicular glia. This phenomenon intervenes only after the axons have reached approximately a 1 or 2 μm diameter. Then the oligodendrocytes send processes toward the nearby axons, and the subsequent development of myelin lamellae involves the rotation of the so-called inner tongue.

As often occurs in science, the establishment of all these facts about myelin has resulted in more questions than answers. For example: What is the origin of the oligodendroglia? What is the signal for their migration? What is the signal for the wrapping around the axon? What is the signal for the end of the wrapping? How is the newly formed myelin deposited? Are the cytoplasmic incisures the real site of maintenance of myelin? Is there any contribution of the axon to the synthesis of myelin? (Indeed, it has been shown in the PNS that radioactive choline injected in the neurons was found in the myelin phosphatidyl-choline.)[12] And also, is there any contribution of myelin to the metabolism of neurons (the presence of carbonic anhydrase in myelin could be interpreted in that way)? How does the regulation of myelin component synthesis occur? How is myelin assembled?

A few of these questions have at least partially been illuminated by studies on dysmyelinated mutant mice as well as the consequences of impaired myelination on other cell types (neurons and astrocytes).

MURINE MUTATIONS AS A TOOL TO STUDY MYELINATION[13,14]

One of the difficulties in the study of CNS myelination at the cellular and molecular levels resides in the fact that it is an integrated process requiring both the axon and the glial cell. Thus, much of the work has to be performed on the intact animal or on tissue cultures. The rat and mouse appear to be particularly useful, since in these species myelination occurs post-natally in brain and so it is possible to follow the whole developmental process during a short period of the animal's life. Although myelination is the most dramatic cytologic event, occurring mainly between day 12 and day 60 after birth in these rodents, other developmental processes occur simultaneously.

Thus, the discovery of mutations in the mouse, whose phenotypic expression mainly involves myelination, has provided a useful tool for studying the neurobiology of myelin for many reasons. On the one hand, these defects, being genetically determined, are stable and stereotyped, which means that they are easily reproducible in large amounts. On the other hand, in a mutant strain, the fate of the animal is predictable, and in many cases investigations can start before the onset of symptoms which always have the same phenotypic expression. This is obviously advantageous in developmental studies, which can even be carried out during embryonic life. In addition, the fact that several mutations affect the same developmental processing at various stages allows dissection of the different steps involved in myelin processing. Actually, in most of these mutants, no causal relationship has been established between the gene defect and the molecular events affecting myelin. Nevertheless, they have been found to provide a useful experimental approach for investigating the sequential events underlying myelin formation and the influence of such events on other developmental processes and on the functions of the nervous system. As in human diseases, the nervous systems of mutants have adjusted to their handicaps, and functional compensatory mechanisms may have developed which can be analyzed using pharmacological tools.

Mutations of the CNS, more or less restricted to the myelination process, present morphological and biochemical modifications which always have, at different degrees, the same phenotypic expression.

From work on the mouse, it is clear that several genes are implicated in myelination, some of which involve, with differing degrees of selectivity, the central and/or the peripheral nervous system. Some of these mutations exhibit clear-cut defects, more or less restricted to the myelination process (quaking, jimpy and its allele jimpy[msd], shiverer and its allele shiverer[mld] affecting mainly CNS, and trembler affecting mainly PNS). Others have a more complex phenotype with associated alterations in the myelin content as a result of the pleiotropic nature of most of the mutations in the vertebrates (twicher, dilute-lethal, wobbler-lethal, crinkled, etc.). We will restrict ourselves to those having a primary effect on myelination in the CNS. The jimpy mutations, discovered by R. Phillips,[15] is recessive, X-linked. The quaking mutant mouse, reported for the first time by R. Sidman,[16] is recessive autosomal (chromosome 17). Shiverer, reported by E. March,[17] is autosomal recessive (linkage not known).

To a first approximation, CNS mutations appear to block specific steps in the myelination program. For jimpy, this is the beginning of myelination and results in the persistence of small axons and immature oligodendrocytes, whereas later steps of oligodendrocyte differentiation are affected in quaking and shiverer. These three mutations involve, to different degrees and at steps, the processing of myelin assembly as shown by the dynamics of myelin specific components, myelin basic protein, and galactosylceramide, from their appearance in the oligodendrocyte to their incorporation into the myelin sheath.

The jimpy mutation as a tool to establish whether or not a given molecule is unique in myelin and oligodendroglia

The extreme severity of dysmyelination in jimpy has been valuable as an index for establishing whether a molecule is a myelin or a non-myelin constituent. Thus, the finding that enzymes such as the cholesterol esterhydrolase or the 2′,3′-nucleotide 3′-phosphodiesterase are depleted in the jimpy is good evidence for their presence in the myelin or the oligodendroglia. Furthermore, now that immunohistochemical techniques are more and more often used, the jimpy mutant can greatly help in the extramyelinic localization of some molecules which are present in the myelin and in other structures. This is well illustrated in the case of the neuronal localization of sulfatide, especially in the brain stem. In the latter structures, most of the neuronal groups are invaded by myelinated bundles. For instance, in the case of the accessory trigemini nucleus, it has been possible only on jimpy tissue sections to evidence the presence of sulfatide on neuronal plasma membranes. Indeed, in normal mice, such a discrete labeling is hindered by the proeminent staining of the neighboring myelin sheaths. Besides this neuronal localization, sulfatide has also been found in extramyelinic glial structures such as ependymal cells or astrocytic feet. The fact that in jimpy these latter sulfatide-positive structures (ependymal cells, astrocytic processes or neurons) are preserved is good evidence for the existence of several pools for sulfatide synthesis, independent from those involved in myelin formation (Fig. 1).

Dysmyelinating mutants as a tool to study the synthesis sequences of myelin constituents

Dysmyelinated mutants are also of great help in determining the sequence of synthesis of myelin constituents and also have partially illuminated their regulation mechanisms. The fact that, in jimpy and in quaking, so many enzyme activities (mainly of the biosynthetic pathway of myelin lipid) have been found reduced while these two mutations are located on different chromosomes suggests a feedback control mechanism resulting from the intracellular accumulation of end products.[13] Indeed, in jimpy, we have found an accumulation of galCb and sulf in the cytoplasm of oligodendrocytes (Fig. 3). Although this was not true in the quaking mutant (Fig. 2), in the shiverer mice (Fig. 4), oligodendroglia were also found to accumulate both galactolipids. In one case (the jimpy), the MBP is synthetized and deposited at a very low level. In the shiverer, there is no immunodetectable MBP which results in the absence of major dense line

section of a 21-day-old shiverer mouse at the level of the ▶ corpus callosum, incubated with antigalactosylceramide antibodies. As in the jimpy section shown on Fig. 3, there is a swelling of oligodendrocytes which are intensely galactosylceramide-positive. Some of the galactosylceramide deposit in the myelin sheaths (arrow), in contrast to what is seen in the jimpy mutant. Technical details have been described in reference 10.

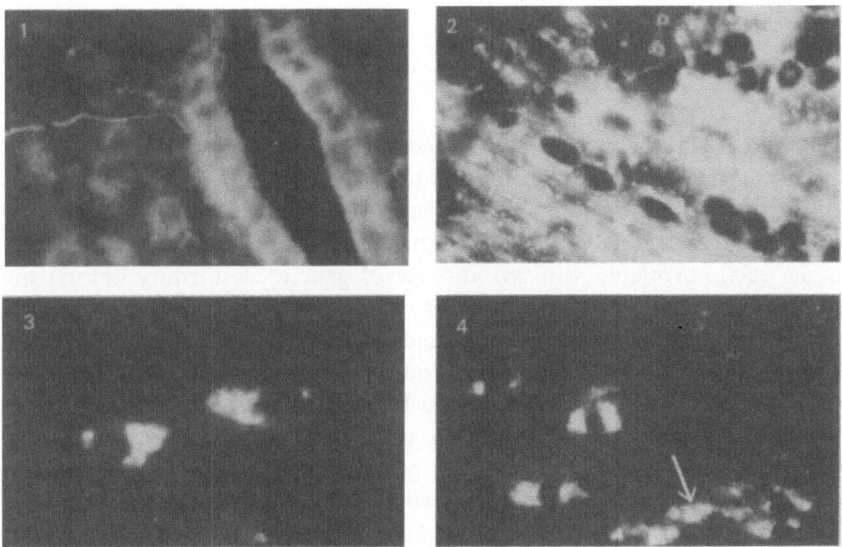

FIG. 1. Indirect immunofluorescence on a coronal brain section of a 21-day-old jimpy mouse at the level of the lateral ventricle, incubated with antisulfatide antibodies. Ependymal cells are sulfatide-positive, and so is one process in contact with them. Technical details have been described in reference 10.

FIG. 2. Indirect immunofluorescence on a coronal brain section of a 30-day-old quaking mouse at the level of the corpus callosum, incubated with antigalactosylceramide antibodies. Most of the fibers are galactosylceramide-positive. Note the fluorescence on the plasma membrane of interfascicular oligoglia. Technical details have been described in reference 10.

FIG. 3. Indirect immunofluorescence on a coronal brain section of a 21-day-old jimpy mouse at the level of the corpus callosum, incubated with antigalactosylceramide antibodies. All the axons are negative, contrasting with the intense fluorescence of the oligodendrocyte. In these cells, galactosylceramide is confined in the cytoplasm which is impressively increased in volume. Technical details have been described in reference 10.

◀ FIG. 4. Indirect immunofluorescence on a coronal brain

(see for details the paper by Mikoshiba *et al.* in this volume). Nevertheless, in both mutants, galactolipids were found to accumulate in the oligodendroglia. This suggests that there is no direct interference of the MBP on the synthesis of galactolipids, and the key regulatory step(s) must be elsewhere.

Cautions related to the use of these mutants. Introduction of genetic markers

During the last ten years, our attention has been drawn to the difficulties in obtaining reproducible and comparative results, when biochemical or immunochemical analysis was performed on dysmyelinating mutants. This is, in part, correlated with an absence of genetic uniformity among the strains or sometimes within a given strain carrying a mutation, as the different mutations have been isolated on different inbred or non-inbred strains. The ideal breeding system would be the maintenance of the mutations on two different inbred strains bred in parallel, the mutants being produced by intercrossing parents of these two strains. For instance, with C57BL/6 $+/qk$ and C3H/He $+/qk$ parents, one would obtain B6C3H qk/qk F1 and $+/qk$ counterparts offering all the precious advantages of the hybrids (heterosis + small variability). The animals thus obtained would be genetically homogeneous and present interesting zootechnical qualities. Unfortunately, this stage has not been reached yet. Also, a common policy concerning the genetic uniformization of the dysmyelinating mutants has not been established among international laboratories working in this field. Nevertheless, these mutations have gross and obvious abnormalities which allow comparisons between them.

Another difficulty is determining whether some of the abnormalities observed in mutants may be directly under genetic control or are merely consequences of the tissue or cellular environment. Thus, a comparison of *in vivo* and *in vitro* studies appears to be extremely useful as it is possible in tissue culture to modify prevailing conditions. Tissue cultures can only be performed on the immature CNS. Also, to determine the primary site of action of a gene, investigations at very early stages of development are necessary. Important events in biology are prepared for in advance of the time at which they actually take place. For many years, research along those lines has been hindered by the fact that the mutations in jimpy and quaking could be detected only when tremors and convulsions appeared (around the 12th day after birth, when the myelin sheath is already detectable by morphological techniques). Also, homozygotes can not breed because of a defect in sperm maturation in the quaking and lethality in the jimpy mutant. Thus, techniques were developed utilizing marker genes as closely linked as possible to the mutated gene, in order to detect mutations at birth or even before.

For quaking, a lethal semi-dominant balanced marker has been introduced: the $T^{2/j}$ mutation located at 2 IU of crossing over from quaking. Double heterozygotes $\left(\dfrac{T^{2/j}\ +}{+\ qk}\right)$ crossed together give lethal homozygotes $\left(\dfrac{T^{2/j}\ +}{T^{2/j}\ +}\right)$, short-tailed double heterozygotes $\left(\dfrac{T^{2/j}\ +}{+\ qk}\right)$ used for breeding, and long-tailed homozygotes $\left(\dfrac{+\ qk}{+\ qk}\right)$. The latter could be detected as early as at the 13th day of embryonic life.

For jimpy, the tabby marker gene has been used. This is located on the jimpy chromosome at 20 units of crossing-over. This means that in 20% of cases the linkage between the marker gene and the jimpy gene will fail to be maintained during meiosis, due to mutual exchange between the chromatids of homologous chromosomes. The presence of a substantial amount of crossing-over is one of the limitations of this marker. The introduction by F. Lachapelle of an inverted segment of the X-chromosome actually suppresses this crossing-over. Heterozygous females $\left(\dfrac{+\qquad Ta\ jp}{In(x)\ 1H\ +\ +}\right)$ crossed with $In(x)$ 1H $+\ +$ males give rise to females of either wild type or heterozygous for $Ta\ jp$, and to males of either wild type or $In(x)$ 1H $Ta\ jp$, expressing both jp and Ta. Thus, jp males can be detected at birth. Under these conditions, the suppression of recombination has been accomplished. In the case of the shiverer mutant it is possible to breed homozygotes, and so these problems do not occur.

Ontogeny of myelin basic protein (Figs. 7–9)

The combination of a very sensitive RIA developed for myelin basic protein (MBP) and of the introduction of marker genes has allowed us to investigate the MBP content of jimpy and quaking in the premyelination period. MBP was quantified using an RIA technique in the spinal cord, cerebellum, brain stem, and cerebral hemispheres. In control animals, it could be detected in the whole brain of 19-day-old fetuses. At birth, MBP was present in every region of the CNS. Its deposition proceeds first at a low rate for several days and then suddenly increases at ages ranging from 1 to 12 days postnatal according to the area. The leveling off of the curves occurs immediately after birth in the spinal cord, at day 5 in the cerebellum

NORMAL

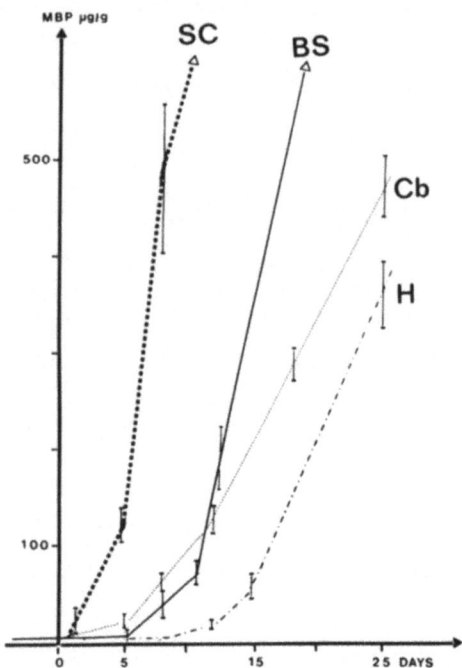

FIG. 7. Appearance and accumulation of MBP in spinal cord (SC), brain stem (BS), cerebellum (Cb), and hemispheres (H) of normal C57BL/6 mice. Values are given in μg MBP per gram of fresh tissue.

and in the brain stem, and around day 12 in the cerebral hemispheres. The accumulation of MBP reaches a maximum of 130 μg/g in the spinal cord between days 5 and 8 postnatal. The maximum daily increment is 3 to 4 times lower in the brain than in the spinal cord.

In quaking mice, the MBP appears first in the spinal cord and last in the hemispheres, following the same caudo-rostral gradient as in normal mice, but major differences can be observed: 1) the date of the appearance of MBP is delayed in quaking compared to normal; 2) the accumulation of MBP is strongly depressed throughout the brain. The final level in the adult varies from 20 % of normal in the brain stem to 5 % in the cerebellum.

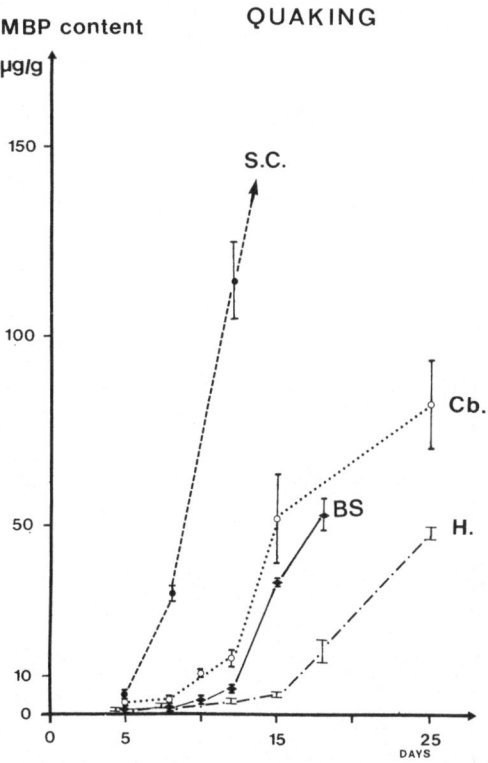

FIG. 8. Appearance and accumulation of MBP in spinal cord (SC), brain stem (BS), cerebellum (Cb), and hemispheres (H) of quaking mice. Values are given in μg MBP per gram of fresh tissue.

In the spinal cord, the MBP deficit tends to decrease even during adulthood, in contrast to the brain where this deficit is maintained during the whole life span. The level of MBP increases from 25% of normal in the spinal cord of a young adult (3 months) to 35% in an old adult (1 year). In that respect, the quaking mutant offers evidence that myelin formation or maintenance can occur in adulthood, at least in the spinal cord.

Although, basic protein is normally synthesized by this mutant, labeling studies[18] have shown that, in the whole brain of a 32-day-old *qk* mutant, MBP is incorporated into myelin at a 35% lower rate than in normal. However, these authors reported that the yield of myelin isolation was only 10% compared to controls. From our curves, it appears that approx-

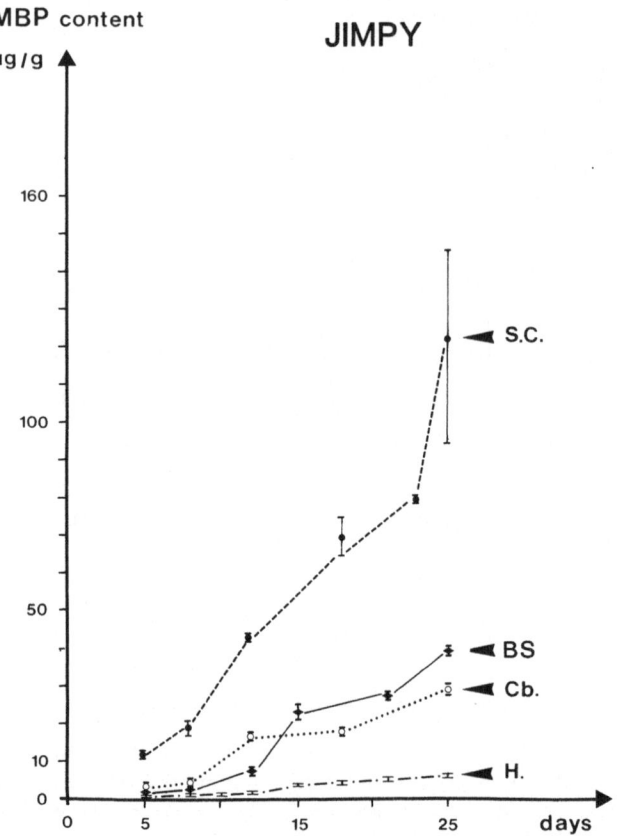

FIG. 9. Appearance and accumulation of MBP in spinal cord (SC), brain stem (BS), cerebellum (Cb), and hemispheres (H) of jimpy mice. Values are given in μg MBP per gram of fresh tissue.

imately 10% of the normal MBP content is present in the whole brain at 32 days. The fact that the final deficit of MBP in *qk* myelin is much higher than its defect of incorporation suggests that MBP is subjected to a degradation process, after having been incorporated into myelin. This could be due either to an accelerated turnover of the protein or to a pathological clearance of fragments of myelin-like material.

In jimpy, the appearance of MBP is already retarded during the premyelination period. Then its deposition is very low, resulting in a deficit ranging from 94% in the cerebellum to 98% in the hemispheres of 25-day-old animals. The daily increment is reduced by 15 to 100 times according to the area. These results are in agreement with the observation by Braun et al.[19] showing that the incorporation of MBP into the myelin sheath is drastically reduced in jimpy while its synthesis is normal.

The observation that MBP is already present in the brain shortly before birth suggests that the process of myelination, although postnatal, is being prepared at a prenatal stage, before any myelin-like material could be morphologically detected. Accumulation of MBP in the oligodendrocytes before its deposition in myelin has already been shown by immunohisto-chemistry.[20,21] The high sensitivity of our RIA technique has allowed the detection of MBP several days before it could be observed by these authors. This implies that, in contrast to what was commonly assumed, the synthesis of myelin components, at least MBP, does not occur imme-diately before myelin deposition starts, but is slowly prepared by the oligodendrocytes. In that respect, during the premyelination period, one should be careful in using MBP as a myelin index. Nevertheless, after this preassembly period, as soon as the wrapping of myelin has started, the amount of MBP incorporated rapidly surpasses the oligodendrocyte content and thus the latter pool becomes negligible. Therefore, one can consider MBP a good index of myelination. The maximum rate of myelin deposition so as the duration of active deposition seems to be characteristic for each region. At the maximum, the amount of myelin deposited per day was calculated to be 4 mg/g of fresh tissue in the spinal cord.

Cell and tissue cultures of mutant brains

Another advantage of the marker gene allowing the early detection of the mutant is the potential to develop tissue and cell cultures. Such an approach should allow the determination of the factors linked to the oligo-dendroglia and the influence of other cell types or of the environment on the mutations.

Brain tissue of jimpy, quaking, and shiverer have already been grown in cultures (reviewed in 14). In organotypic cultures, it is possible to cor-relate the deficiencies found in vivo. Such a finding rules out the putative involvement of extrinsic factors. Dissociated brain cell cultures of mutants have also been performed. Using such an approach, Sandru et al.[22,23] were able to show that, in vitro as in vivo, jimpy oligodendroglia were able to express galactosylceramide. These authors found also that the expression of MBP in jimpy culture was strongly depressed when compared to controls. It must be stressed that, in their paradigm (i.e., dissociated brain cell cultures), MBP was expressed seven days after galactosylceramide and only 0.1% to 1.5% of galactosylceramide-positive cells were found to be MBP-positive. The delayed apparition of MBP compared to galactosylceramide in dissociated culture has been reported by other authors. This is in contrast with the *in vivo* situation where it has been shown that MBP is expressed far in advance from galactosylceramide. This raises the question of what type of culture should be used (as organotypic cultures seem to mimic more the *in vivo* situation) and also at what age the brains should be taken to be put in culture (in the embryonic stage, the oligodendroglia is less differentiated and thus more suitable for study in culture).

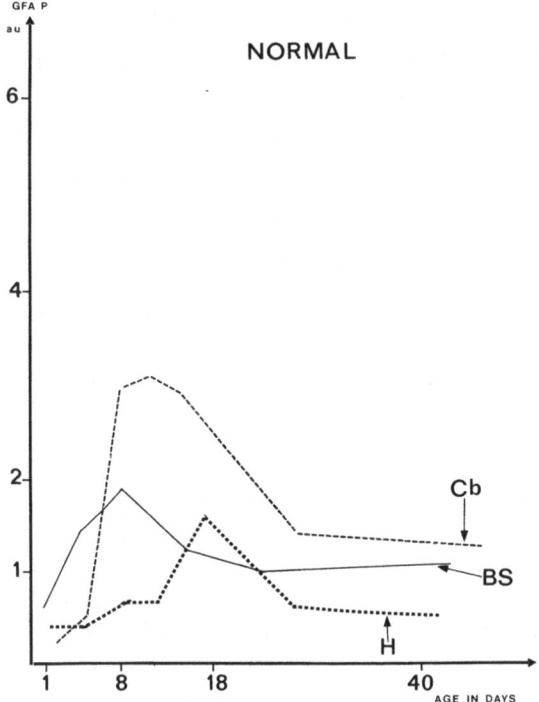

FIG. 10.   Accumulation of GFA protein in brain stem (BS), cerebellum (Cb), and hemispheres (H) of normal C57BL/6 mice. Values are given in arbitrary units (au).

Dysmyelinating mutants as a tool for the study of neuron-glia interactions
Myelin-deficient mutants are characterized by a clinical phenotype (i.e., tremors and convulsions) and a "morphological" phenotype (i.e., dysmyelination). Do these phenotypic expressions, which allow the detection of the mutants, necessarily relate to dysmyelination? As stressed by Norton,[24] "there are mutants which reel and wobble but which don't have myelin abnormalities and there may well be mutants with abnormal myelin which don't shiver, tremble, jimp or quake and therefore would not be detected." This raises the question of the influence of an ontogenic defect on the development of other cellular structures or on structures already differen-

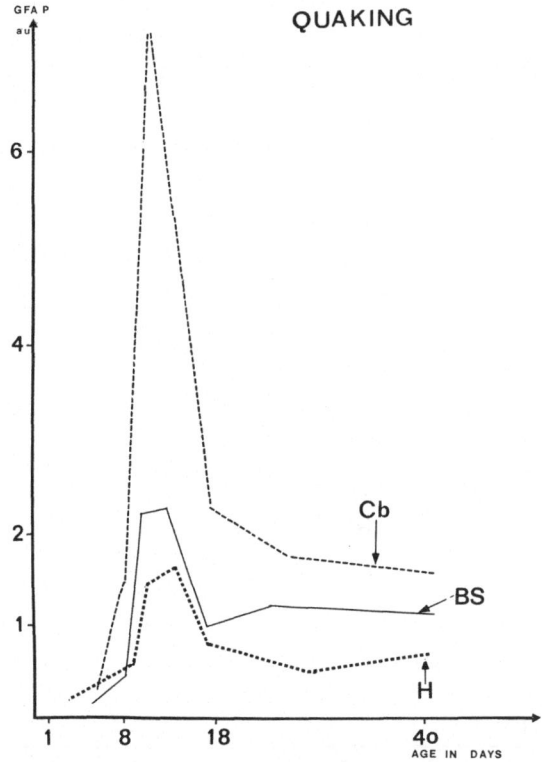

FIG. 11. Accumulation of GFA protein in brain stem (BS), cerebellum (Cb), and hemispheres (H) of quaking mice. Values are given in arbitrary units (au).

tiated. We have undertaken the study of two aspects of this problem. One is concerned with the astrocyte-neuronal interactions which exist in these mutants, the other with the influence of myelination impairment on neuronal function.

GFAP as an index of astrocyte activity in the mutants (Fig. 10–12)

The glial fibrillary acidic protein (GFAP) was measured using rocket immunoelectrophoresis. It could be detected at birth in every area of the central nervous system of normal C57 B1/6J mice. A steep increase follows, reaching a maximum for each area concomitant with the burst of myelin

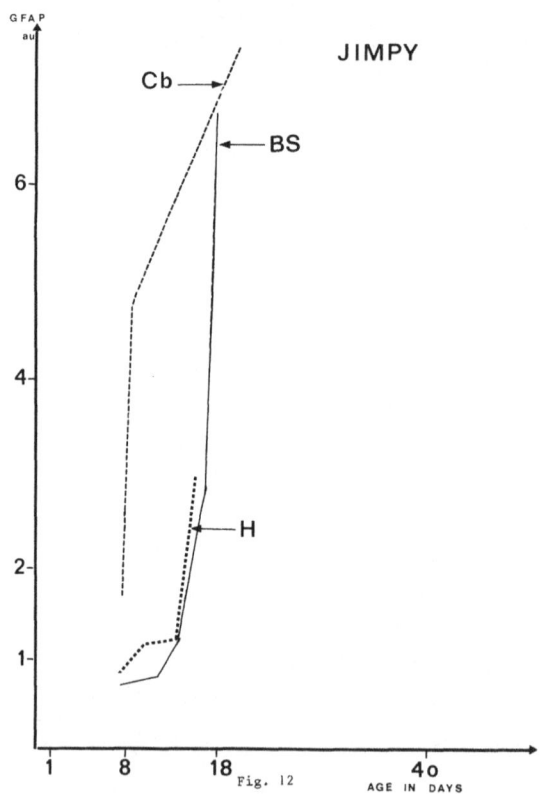

FIG. 12. Accumulation of GFA protein in brain stem (BS), cerebellum (Cb), and hemispheres (H) of jimpy mice. Values are given in arbitrary units (au).

synthesis. This maximum occurs postnatally around day 8 in the spinal cord, day 12 in the cerebellum and in the brain stem, and day 15 in the hemispheres. Then the amount of GFAP decreases, relative to total proteins, and reaches a plateau more or less rapidly according to the area. The sequence of accumulation of GFAP is not significantly modified in the CNS of quaking mutants. However, the average which follows the onset of myelinogenesis occurs normally. Interestingly, in the spinal cord, the GFA protein follows a quite different curve. It keeps accumulating after that point so that, in the adult, its concentration is 10 to 15 times higher than in the spinal cord of controls.

In the CNS of jimpy mutants, the developmental curve of the GFAP is slightly delayed to older ages. Moreover, the accumulation proceeds rapidly and continuously until death. As seen by immunofluorescence on jimpy brain tissue sections, in addition to their increased number, astrocyte processes are different when compared to normals in that they seem to be oriented in the same direction as the axonal track (Fig. 6). In contrast, in dissociated brain cell cultures where no myelin is formed, the number and morphology of jimpy astrocytes has been found normal. This strongly suggests that, in jimpy, gliosis is a reactive phenomenon of defense.

Pharmacological approach to neuronal dysfunction in the mutants

Extremely complex interactions occur between neurons and glia at every stage of myelinogenesis. Considering dysmyelinating mutants, effects on neurons are probably more related to functional interactions than to the pleiotropic effect of the mutations. Abnormal gait, tremor, and convulsions are only evidenced at the 12th day when there begins to be a patent lack of myelin, although in mouse brain, neurons have already been formed before birth and synaptogenesis is already partly developed at this stage.

As in some human genetic and degenerative disorders, morphological alterations have been demonstrated by Friedrich[25] in the Purkinje cell axons of quaking and shiverer mutations. There are peculiar clusters on densely packed small vesicles in several terminals associated with axonal focal swellings. These manifestations possibly represent secondary reactions to the lack of myelin which could be related to the abnormal firing pattern or to disrupted trophic interactions.

In jimpy mutants, no major changes in neuronal size, number, or arrangement have been described, and axons seem normal although their growth has been shown to be retarded.

Thus, the role of neurons in the expression of clinical symptoms of these

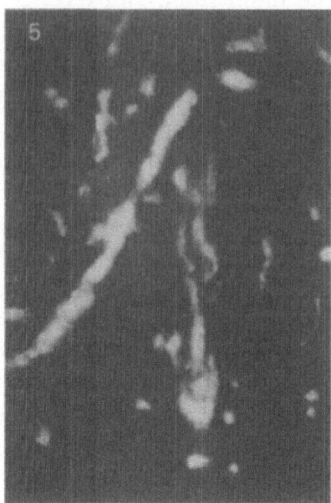

FIG. 5. Indirect immunofluorescence on a coronal brain section of a 21-day-old jimpy mouse at the level of the corpus callosum, incubated with anti-MBP antiserum diluted 1/30. No oligodendrocyte is seen, but a few myelin sheaths are clearly seen.

FIG. 6. Indirect immunofluorescence on a coronal brain section adjacent to that in Fig. 5, incubated with anti-GFAP antiserum diluted 1/50. Most of the astrocytic processes are disposed in the same direction as the axons.

mutations is still poorly understood. A new field in the study of the functional defect of these mutations is now opening by the use of pharmacological tools.

Abnormalities of neurotransmitters in quaking brains have been reported by several authors, especially an increase in cerebral dopamine, noradrenaline and acetylcholine.

One of the questions which can be asked is whether the clinical symptoms observed in mutants are related to a general trophic defect in neurons or if the defects in myelin content observed lead to functional alterations of specific systems. To answer this question, we have chosen a pharmacological approach to the convulsions which occur in the quaking mice and which may be similar to those observed in the late stages of human leukodystrophies.

Drugs affecting cholinergic and serotoninergic metabolism are without effect on the convulsions of the quaking mouse. Several drugs activating the noradrenergic system prevent convulsions, among them direct alpha-1 receptor agonists (norepinephrine, phenylephrine) as well as drugs which increase physiological release of norepinephrine (e.g. yohimbine) via the blockade of alpha-2 receptors. The anticonvulsant effect of yohimbine[26] is due to the blockade of presynaptic alpha-receptors. This is suggested by the fact that it can be blocked by both clonidine (presynaptic alpha agonist) and prazosine (postsynaptic alpha blocker). Beta-adrenergic systems do not seem to be involved.

It is striking that the effect of drugs modulating the alpha-receptors is exactly the opposite of that seen in the model proposed by Horton et al.[27] Working with audiogenic seizures in the DBA/2 mouse, these authors found a protective effect of clonidine with a certain degree of potentiation by yohimbine, thus indicating that the mechanisms of convulsions which occur in severe dysmyelinating diseases are different although involving also noradrenergic systems.

In vitro binding studies support the in vivo pharmacological results since alpha-adrenoreceptors appear to be modified in the brains of the quaking mice. There is an increase in the number of alpha-adrenoreceptors, which may play a role in the behavioral abnormalities observed in this mutant.

Although we are still far from explaining the consequences of myelin defects on neuronal dysfunction, this functional approach indicates that alterations in specific systems involving catecholamines exist.

CONCLUSION

In this review, we have developed only a few aspects of the contribution of myelin-deficient mutants to the understanding of such a complex and integrated developmental process as myelinogenesis. We have restricted ourselves to the CNS and thus have not debated the trembler mouse, which, in some ways, is the counterpart of the jimpy in the peripheral nervous system. Studies on CNS mutations affecting myelination started some 15 years ago and were underlined by two main ideas: the hope that mutants would serve as models for human genetic diseases and (by analogy to studies on procaryotes) the understanding of normal regulatory mechanisms by analyzing a pertubated system. The fact is that these hopes have so far been only poorly fulfilled. Except for the twitcher mouse,[28,29] which has been shown to be a model for Krabbe's disease, none of the other mutants has a clear correlation with a human disorder. Furthermore, in the twitcher, the genetic defect has been defined. As in human Krabbe's disease, the twitcher mice are defective in galactosylceramidase.

There is also strong evidence that the genetic defect for the shiverer is concerned either with the lack of synthesis or the synthesis of a defective MBP. Further studies on this mutant should now allow a better understanding of the role of MBP on the compaction of myelin.

Quaking and jimpy have been the most studied mutations. Nevertheless, the search for their respective genetic defects has not been successful. After so many years, one might be tempted to give up. This is an understandable and maybe even a wise attitude, but it is also a pessimistic view of the situation. An optimistic way of thinking is to consider that these mutations are concerned with key regulatory mechanisms of myelination and thus are worthy of some additional effort. Our past failure was due to inadequate approaches and techniques. New tools such as immunohistochemistry, cell type markers, and tissue culture have only quite recently been developed. They have now advanced to a point where rapid progress can reasonably be expected.

Furthermore, pharmacological approaches, where these mutations are considered as a paradigm for studying neuro-glia interactions, have already given promising results. Thus, our bet on the future is that the optimistic attitude is not a utopian one.

REFERENCES

1. Morell, P. and Norton, W. T.: *Myelin*, **242**: 88–118. Scientific American, 1980.
2. Morell, P. (ed): Myelin. Plenum Press, New York, 1977.
3. Peters, A., Palay, S. L. and Webster H. de F. (eds.): The Fine Structure of the Nervous System: The Neurons and Supporting Cells. W. B. Saunders, New York, 1976.
4. Mugnaini, E.: Fine structure of myelin sheaths. In: Proceedings of the European Society for Neurochemistry, Vol. 1 (ed. V. Neuhoff), pp. 3–31. Verlag Chemie, Weinheim, 1978.
5. Baumann, N.: Biosynthesis and deposition of myelin lipids. In: Proceedings of the European Society for Neurochemistry, Vol. 1 (ed. V. Neuhoff), pp. 48–63. Verlag Chemie, Weinheim, 1978.
6. Waehneldt, T. V.: Proteins of central nervous system myelin. In: Proceedings of the European Society for Neurochemistry, Vol. 1 (ed. V. Neuhoff), pp. 32–47. Verlag Chemie, Weinheim, 1978.
7. Matthieu, J. M.: Biosynthesis of myelin proteins and glycoproteins. In: Neurological Mutations Affecting Myelination. INSERM Symposium No.

14 (ed. N. Baumann), pp. 275–298. Elsevier/North-Holland, Amsterdam, 1980.

8. Joffe, S., Rapport, M. M. and Graf, L.: Identification of an organ specific lipid hapten in brain. *Nature* (London), **197**: 60–61, 1963.

9. Raff, M. C., Mirsky, R., Fields, K. L., Lisak, R. P., Dorfman, S. H., Silderberg, D. H., Gregson, N. A., Leibowitz, S. and Kennedy, M.: Galactocerebroside: A specific cell surface antigen marker for oligodendrocytes in culture. *Nature* (London), **274**: 813–816, 1978.

10. Zalc, B., Monge, M., Dupouey, P., Hauw, J. J. and Baumann, N. A.: Immunohistochemical localization of galactosyl and sulfogalactosyl ceramide in the brain of the 30 day-old mouse. *Brain Res.*, **211**: 341–354, 1981.

11. Barbarese, E., Carson, J. H. and Braun, P. E.: Identification of prelarge and presmall basic proteins in mouse myelin and their structural relationship to large and small basic proteins. *Proc. Nat. Acad. Sci. U.S.A.*, **74**: 3360–3364, 1977.

12. Droz, B., Di Giamberardino, L., Koenig, H. L., Boyenval, J. and Hassig, R.: Axon-myelin transfer to phospholipid components in the course of their axonal transport as visualized by radioautography. *Brain Res.*, **155**: 347–353, 1978.

13. Hogan E. L.: Animal models of genetic disorders of myelin. In: Myelin (ed. P. Morell), pp. 489–520. Plenum Press, New York, 1977.

14. Baumann, N. (ed.): Neurological Mutations Affecting Myelination. INSERM Symposium No. 14. Elsevier/North-Holland, Amsterdam, 1980.

15. Philipps, R. J. S.: Jimpy: A new totally sex linked gene in the house mouse. *Z. indukt. Abstammungs-Vererbungsl.*, **86**: 322, 1954.

16. Sidman, R. L., Dickie, M. M. and Appel, S. H.: Mutant mice (quaking and jimpy) with deficient myelination in the central nervous system. *Science,* **144**: 309, 1964.

17. March, E.: Personal Communication. *Mouse News Letter*, **48**: 24, 1973.

18. Brostoff, S. W., Greenfield, S. and Hogan, E. L.: The differentiation of synthesis from incorporation of basic protein in quaking mutant mouse myelin. *Brain Res.*, **120**: 517–520, 1977.

19. Braun, P. E., Pereyra, P. M. and Greenfield, S.: Mechanisms of assembly of myelin in mice: A new approach to the problem. In: Neurological Mutations Affecting Myelination. INSERM Symposium No. 14 (ed. N. Baumann), pp. 413–422. Elsevier/North-Holland, Amsterdam, 1980.

20. Sternberger, N. H., Itoyama, Y., Kies, M. W. and Webster, H. de F.: Myelin basic protein demonstrated immunochemically in oligodendroglia prior to myelin sheath formation. *Proc. Nat. Acad. Sci. U.S.A.*, **75**: 2521–2524, 1978.

21. Hartman, T., Agrawal, H., Kalmbach, S. and Shearer, W.: A comparative study of the immunohistochemical localization of basic protein to myelin and oligodendrocytes in rat and chicken brain. *J. Comp. Neurol.*, **188**: 273–290, 1979.

22. Sandru, L., Siegrist, H. P., Wiesmann, U. N. and Herschkowitz, N.: Development of oligodendrocytes in jimpy brain cultures. In: Neurological Mutations Affecting Myelination. INSERM Symposium No. 14 (ed. N. Baumann), pp. 469–474. Elsevier/North-Holland, Amsterdam, 1980.
23. Bologa-Sandru, L., Zalc, B., Herschkowitz, N. and Baumann, N.: Oligodendrocytes of jimpy mice express galactosylceramide: An immunofluorescence study on brain sections and dissociated brain cell cultures. *Brain Res.*, **225**: 425–430, 1981.
24. Norton, W. T.: Concluding remarks. In: Neurological Mutations Affecting Myelination. INSERM Symposium No. 14 (ed. N. Baumann), pp. 545–560. Elsevier/North-Holland, Amsterdam, 1980.
25. Friedrich, V. L., Donna, J. R., Konteckt, L. and Massa, P. T.: Neuronal abnormalities in the cerebellum of quaking and shiverer mice. In: Neurological Mutations Affecting Myelination. INSERM Symposium No. 14 (ed. N. Baumann), pp. 141–146. Elsevier/North-Holland, Amsterdam, 1980.
26. Chermat, R., Lachapelle, F., Baumann, N. and Simon, P.: Anticonvulsant effect of yohimbine in quaking mice: Antagonism by clonidine and prazosine. *Life Science*, **25**: 1471–1476, 1979.
27. Horton, R., Anlezark, G. and Meldrum, B.: Noradrenergic influences on sound-induced seizures. *J. Pharmacol. Exp. Ther.*, **214**: 437–442, 1980.
28. Duchen, L. W., Eicher, E. M., Jacobs, J. M., Scaravilli, F. and Teixeira, F.: A globoid cell type of leukodystrophy in the mouse: The mutant twitcher. In: Neurological Mutations Affecting Myelination. INSERM Symposium No. 14 (ed. N. Baumann), pp. 107–114. Elsevier/North-Holland, Amsterdam, 1980.
29. Kobayashi, T., Scaravilli, F. and Suzuki, K.: Biochemistry of twitcher mouse: An authentic murine model of human globoid cell leukodystrophy. In: Neurological Mutations Affecting Myelination. INSERM Symposium No. 14 (ed. N. Baumann), pp. 253–256. Elsevier/North-Holland, Amsterdam, 1980.

Discussion

Dr. Mullen: I don't entirely agree with Dr. Zalc that our reason for using mutants is to understand normal physiology. If everyone were normal we would have difficulty finding support for our research. I agree that the twitcher is perhaps the only case where *homology* with a human condition has been demonstrated, but I think many of the other mutants are *analogous* to human conditions.

Studies on Genetic Control of Myelinogenesis in Tissue Culture from Rat and Mouse Cerebella

Kazuhiro Nagaike and Yasuzo Tsukada

ABSTRACT

In the cultures of normal rat cerebella from the 14th, 16th, and 18th days of gestation, the day of birth, and the 2nd day of postnatal life, myelination in each case started on days 16, 13, 11, 9, 8, after incubation, respectively. Accordingly, it seemed that the onset of myelination was determined by chronological age but not on incubation period.

When the cultures from the 18th gestation day to the 2nd day of postnatal life were exposed to an inhibitor of DNA synthesis, myelin formation was not affected compared to the control cultures in spite of the decrease of external granule cells and astrocytes. However, when an inhibitor of DNA synthesis was added to the culture from the 16th gestation day, myelinogenesis was completely inhibited. From these results, it was assumed that proliferation of myelin producing cells in rodent cerebellum occurred between the 16th and 18th days of gestation, while myelin was formed postnatally without granule cells and astrocytes.

In cultured cerebellar tissue from newborn quaking and shiverer mice, myelination was scarcely observed, although the astrocytes in the outgrowth region and both the neuronal cell body and axon looked normal. The addition of the conditioned culture medium from quaking to the control culture had no effect on myelinogenesis.

The activity of CNPase, which is known to be one of the myelin components, in the explants from shiverer cerebella increased with the days of incubation as in the control. From these observations, it was suggested that dysmyelination in the quaking or shiverer can be attributed to certain genetic defects in the oligodendrocytes.

INTRODUCTION

It is well known that the myelin sheath is formed from the process of glial cell cytoplasmic membranes and plays an important part in the expression of neuronal function. Myelin formation appears simultaneously

Department of Physiology, School of Medicine, Keio University, Tokyo, Japan.

in each brain region. Myelinogenesis is thought to be affected by factors such as malnutrition, hormonal deficiency, and amino acid imbalances. The mechanism of myelinogenesis has been studied by many investigators. An important problem to be determined is whether myelinogenesis is under genetic or epigenetic control.

Cultures of neural tissue have been used to study the mechanism of cell differentiation in the nervous system. Tissue culture has an advantage in that the tissue is isolated and placed in a controlled environment which can be easily manipulated.

Myelination in cultured tissue was first reported in 1955 by Peterson and Murray,[1] and Hild[2] in 1957 succeeded in myelin formation in cultured tissue from the kitten cerebellum. Light and electron microscopic examinations of living and fixed tissues from newborn cultured rodent cerebella revealed a typical population of neurons and glial cells, and the formation of myelin and synapses.[3-9]

Dysmyelinating mutant mice were thought to be useful tools for studies of myelinogenesis and neurological disorder. The homozygotes of quaking and shiverer mutants having autosomal resessive mutations are characterized by a severe myelin deficiency in the central nervous system. There have been many reports of morphological and biochemical studies of these mutants *in vivo*.[10-20] However, it is not yet clear whether the cause of dysmyelination in the mutant mice is due to genetic defects in oligodendrocytes or some epigenetic factors such as hypertrophy of the astrocyte or the existence of a toxic substance. In order to clarify the problems mentioned above, here we report on myelinogenesis using several kinds of tissue cultures from rodent cerebella.

MATERIALS AND METHODS

Animals

The rodent cerebellar cultures were prepared from *ddY* mice during postnatal life, from Wistar rats of prenatal age, and from newborn mutant mice. Homozygous $(qk+/qk+)$ and heterozygous $(qk+/+T)$ quaking mice were obtained by mating a pair of animals heterozygous for both quaking and for T, and homozygous shiverer mice were obtained by mating a pair of homozygous shiverer mice. These mice were inbred in our laboratory.

Tissue culture (Fig. 1)

In most culture experiments, each cerebellum of the newborn mice, after removal of the meninges and choroid plexus, was divided by parasagittal section into 12 explants of approximately equal size. In the other culture, the cerebella from rats and mice were dealt with in the same way and sliced into 0.2 to 0.3 mm thick parasagittal sections. The fragments were explanted on a collagen-coated coverslip and maintained in a Maximow double coverslip assembly according to the method of Bornstein and Murray.[3,21] The culture medium used was as follows: one medium was composed of equal parts of Gey's BSS, Eagle's MEM, fetal calf serum, 9-day-old chick embryo extract supplement with final concentration of 600 mg% glucose; and another was 25% fetal calf serum containing

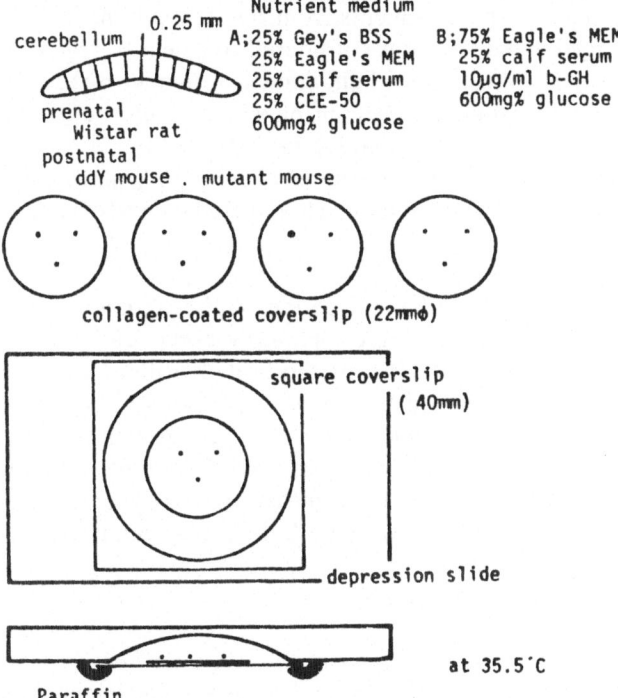

FIG. 1.   Preparation and maintenance of the cultures. Explants were prepared from newborn *ddY* mice, from the embryos of Wistar rats and newborn mutant mice. The fragments were explanted on collagen coated coverslips and maintained in a Maximow double coverslip assembly according to the method of Bornstein and Murray. See text for detail.

Eagle's MEM supplement with final concentrations of 600 mg% glucose and 10 μg/ml bovine growth hormone (Miles, Lot 12). The medium made no difference in the culture morphologically and biochemically. The cultures were incubated at 35.5° C and the medium was changed every three days.

Observation of myelination

Cultures were observed daily under a bright field microscope with a long-working distance oil immersion object and phase contrast microscopy. The rate of myelination is expressed as follows:

$$\frac{\text{Number of myelinated fragments from one cerebellum}}{\text{Number of explanted fragments from one cerebellum}} \times 100 \ (\%)$$

A single myelinated fiber was considered sufficient to regard a fragment as myelinated. The material for light and electron microscopic observation was fixed with 5% glutaraldehyde in phosphate buffer, and then fixed in 2% osmium tetroxide and embedded in Epon 812 or Supper. For microscopic observation, the culture was fixed with 10% phosphate buffered formalin and stained with sudan black and Bodian's silver impregnation.

Biochemical assay

The activity of 2',3'-cyclic nucleotide 3'-phosphohydrolase (CNPase) was assayed with potassium adenosine-2',3'-cyclic phosphate as a substrate in accordance with the method of Kurihara and Tsukada.[22] Nucleic acids were extracted by the procedure of Lowry, and DNA was measured by the method of Kissan et al.[23] DNA synthesis was assayed by the measurement of the amount of 6-^{3}H-thymidine incorporated into DNA fraction. Protein content was measured according to the method of Lowry et al.,[24] using bovine serum albumin as a standard.

RESULTS AND DISCUSSION

Onset of myelin formation on newborn mouse cerebellum

Cerebellar cultures from newborn mice are well documented morphologically by light and electron microscopic techniques. We also observed myelination and differentiation of neurons in culture from a newborn mouse cerebellum (Fig. 2). In this culture, myelination started between 9 and 13 days when they were maintained in two different media. Myelination was never observed before 8 days in culture (Fig. 3). Allerand and Murray[25] reported that cultures of newborn albino Pariss RIII mice commenced myelination from 5 days in culture, but Kies et al.[26] reported

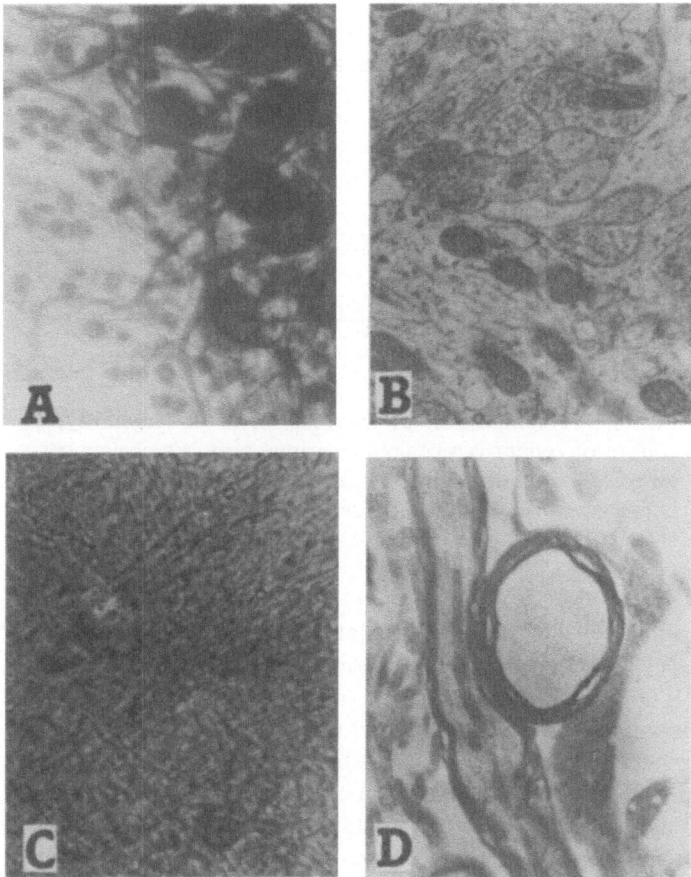

FIG. 2.   Cerebellar culture from a newborn *ddY* mouse. A;
Mature Purkinje cells, 26 days in culture, Bodian silver
impregnation. B; Synapses (large dendrite of Purkinje cell),
12 days in culture. C, D; Myelinated axons, 14 days in culture,
observed in the living state and under an electron microscope.

that myelination of Swiss-Webster mice began between 9 and 12 days in
culture. In our culture system, heterozygous quaking mice, maintained
in an inbred strain of C57BL background, formed myelin from 6 days in
culture. It seems that the onset of myelination in culture of cerebellum
which is devoid of extracerebellar afferents, and mossy and climbing

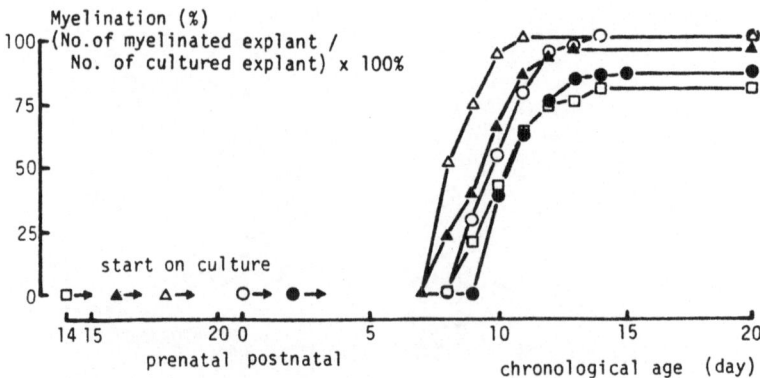

FIG. 3. A comparison of myelination in an organo-typic
culture of rodent cerebellar at various ages. Cerebellar cultures were obtained from Wistar rats 14 (□), 16 (▲), and 18
(△) days embryo, and from mice 0 (○) and 2 (●) days postnatal life. The data are plotted on the same scale as tissue age
in vivo.

fibers, might be determined by genetically programed processes of oligodendrocytes but not by the culture conditions.

Myelinogenesis on cerebellar culture at various ages

Differences in the tissue cultures between prenatal and postnatal life
have been observed in the extent of the outgrowth zone during the first
several days in culture. One or two days after explantation, the outgrowth
from the newborn culture proceeded with a separated bud which rapidly
converged to form a dense sheath of cells. Explants of rat cerebellum from
the 14th and 16th gestation days extended neurites without cell outgrowth
at the beginning. Cell outgrowth from the explants tended to appear earlier
during incubation, as the explants prepared were chronologically older.
In the culture from prenatal rats, myelin formation could be observed as
well as the culture from newborn rats.

Usually, the axons of Purkinje cells from Wistar rats and *ddY* mice were
myelinated at the 8th or 9th day of postnatal life. The time of onset of
myelin formation was investigated in the cerebellar cultures of *ddY* mice,
newborn and on the 2nd day of postnatal life, and of Wistar rats from the
14th, 16th, and 18th gestation days (Fig. 3). When the cerebella of the 14th,
16th, and 18th gestation days and the 2nd postnatal day were explanted,

myelination started at 16 days, 13 days, 11 days, and 8 days in culture, respectively. These periods of time corresponded to the 8th to 10th day of postnatal life. Even though the time of the plantation of the cerebellar tissues varied from the 14th gestation day to the 2nd postnatal day, the onset of myelin formation was the 8th to 10th day of postnatal life having no connection with icubation time. The time schedule of myelination in culture was identical with that *in vivo*. Consequently, it was suggested that the onset of myelinogenesis in rodent cerebella could be determined genetically but not by humoral factors during incubation.

Effects of antimitotic drugs

It has been reported that the neurogenesis and gliogenesis of the rodent cerebellum proceeded on and after birth,[27-30] and the oligodendrocyte is known to play a role in myelination in the central nervous system. However, it is not yet clear when myelin producing oligodendroglia have a spurt of proliferation in rodent cerebella. As for DNA synthesis in

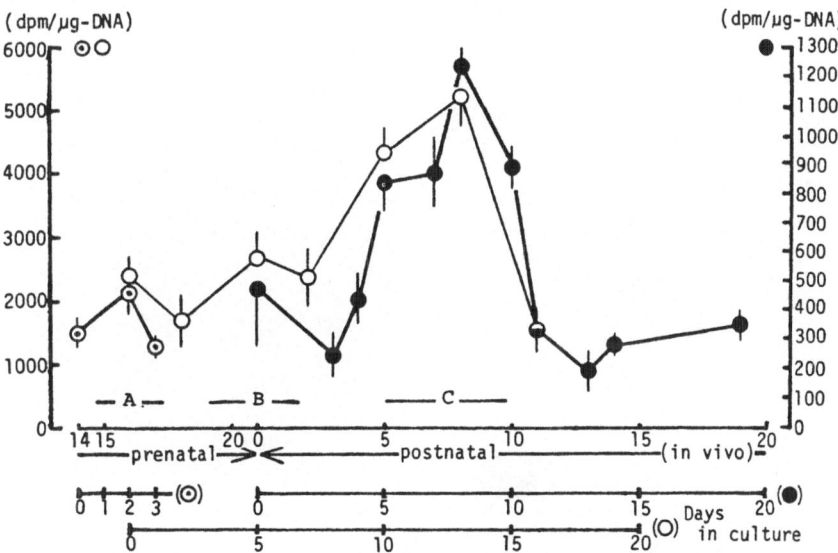

FIG. 4.   Developmental changes of DNA synthesis in rodent cerebellar culture. Each point for dpm/μg-DNA represents the mean ± S.D. by the demonstration of incorporation of ³H-thymidine into DNA fraction.

cultured cerebella, three peaks were observed during development (Fig. 4). The first peak appeared at 16 days of embryonal age (A), the second appeared just after birth (B), and the third corresponded to the 8th day of postnatal age (C).

In order to know which peaks do indicate the proliferation of oligodendrocytes, a DNA synthesis inhibitor, such as methylazoxymethanol (MAM) or bromodeoxyuridine (BUdR), was added to the culture media. When the cerebellar cultures from the 18th gestation day were exposed to the inhibitor of DNA synthesis from the beginning, the outgrowth of the cells appeared without delay but the number of migrating cells was drastically reduced compared to the control (Fig. 5). The effect of the inhibitor was attributed to the absence of proliferation on external granular cell and glial cell precursors. However, the onset of myelination did not differ between the control and the treated cultured cerebella after the 18th gestational day (Fig. 6). The number of myelinated fibers in MAM-treated culture was elevated compared to that of the control culture.

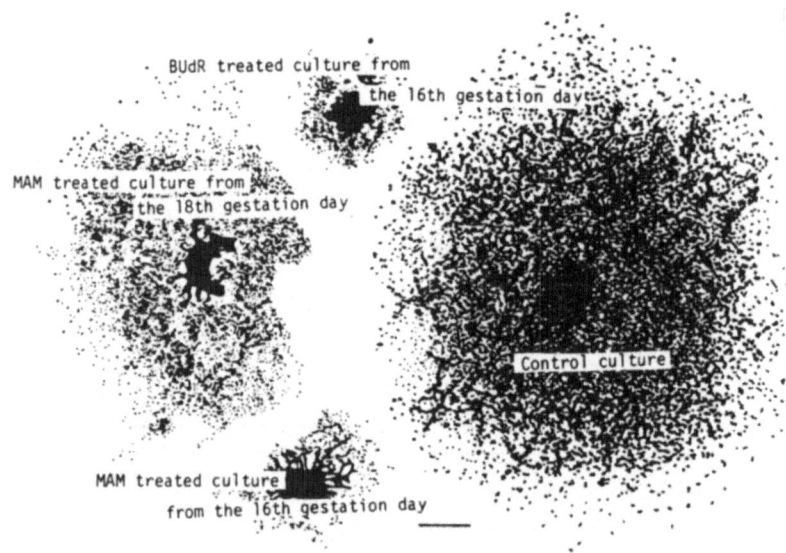

FIG. 5. Effect of an anti-mitotic drug on cell outgrowth from cerebellar explants. A dot indicates one cell in the outgrowth region and the center block in each figure is the explant. (bar = 1 mm)

On the other hand, it was observed that the addition of the inhibitor to the culture from the 16th gestational day had the effect of complete inhibition of cell migration from the explants and of marked reduction of myelination in the explants (Fig. 6). The explants of this culture contained

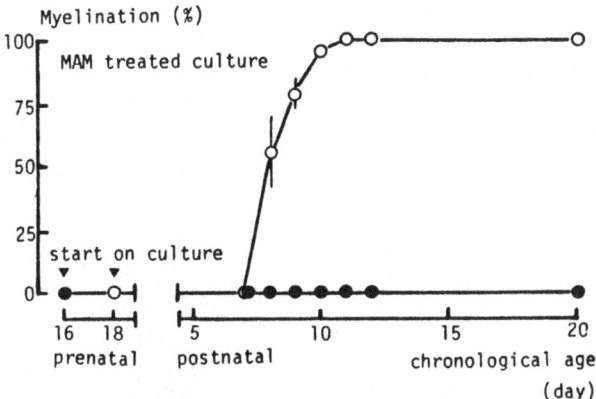

FIG. 6.   The effect of an anti-mitotic drug on myelination in the cerebellar culture from the 16th (●) and the 18th (○) gestation days. Both cultured cerebella were exposed by 20 μg/ ml methylazoxymethanol from the beginning of incubation.

large neurons, probably Purkinje cells, and vacuoles having astroglial and granule cell soma of equivalent size. Complete inhibition of myelinogenesis was observed following the addition of MAM to the medium from the 16th to 20th day of gestation. Privat and Drian[31] reported that myelin formation on cultured tissue with MAM was increased, but Silberberg *et al.*[32,33] observed an inhibition of the myelin formation with the addition of BUdR or cytosine arabinocide. In our experiments, inhibition of myelinogenesis occurred only in the presence of antimitotic drugs in culture from the 16th day of gestation. Then, it was assumed that the effect of the antimitotic drug depended on the time when it was added. From *in vivo* observations by Skoff *et al.*,[34,35] it was reported that oligodendrocytes in rat optic nerves appear on the 5th day of postnatal life. Mares and Bunckner[36] showed that maximum proliferation of oligodendrocytes occurs on the 7th day in the rat visual cortex. Myelination in this region began after oligogenesis. From our experiment, it was considered that proliferation of myelin-producing cells in the cerebellum occurred prenatally, while myelinogenesis appeared postnatally.

It is well known that granule cells and astrocytes of rodent cerebella porliferate postnatally, and the former make synapses to the Purkinje cell dendrites *in vivo* and *in vitro*. However, in culture, the antimitotic agent had no effect on myelin formation when it was added after the 18th gestation day. These findings showed that granule cells and astrocytes had no relation to myelinogenesis in the cerebellar tissue.

From these results, it was tentatively concluded that the onset of myelination in cerebellar tissues of the *ddY* mouse and Wistar rat in the presence of axons to be myelinated was determined by genetic factors in the oligodendrocytes without epigenetic factors, such as electrical activities in the Purkinje cells or the number of astrocytes and granule cells in the cerebellum (Fig. 7).

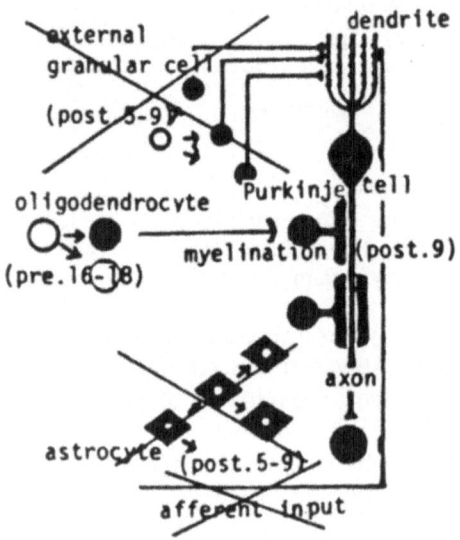

FIG. 7. Schematic representation of the factors affecting myelination in the cultured cerebellum. Since afferent inputs such as mossy and climbing fibers are eliminated by the plantation procedure, these inputs should be eliminated from the effect on myelination. From the experiment using DNA synthesis inhibitor, granule cells or astrocytes dividing postnatally would have no effect on the myelination. It is suggested, therefore, that the oligodendrocytes, which probably proliferate at the 16th to 18th gestation day, are responsible for the myelination in the cultured cerebellum.

Culture of dysmyelinating mutant mouse cerebellum

In order to analyze the factors that govern myelinogenesis in the mammalian nervous system, two mutant mice, quaking and shiverer, were investigated in the culture system.

Quaking in vitro

To obtain myelination of mouse cerebellum *in vitro*, the explants must be taken before the 2nd day of life. Since a clear distinction could be drawn between the homozygous ($qk+/qk+$) and heterozygous ($qk+/+T$) quaking mutant mice used by the length of their tails, we could easily distinguish homo- and heterozygous mice indicating the phenotype of T (Fig. 8). Then, the cerebellum from the homozygous mice were explanted within 24 hours after birth. For a few days after explantation no difference in any feature was observed among cultures from homozygous, heterozygous, and control mice (Fig. 9). The abundance, state of health, and spectrum of shapes in the outgrowth cells were similar in all cultures. In cultures from normal and heterozygous mice, myelin sheaths appeared in many explants after the 10th day. However, myelin was never found in the cultures from quaking mice, with one exception (Fig. 10). By electronmicroscopic obser-

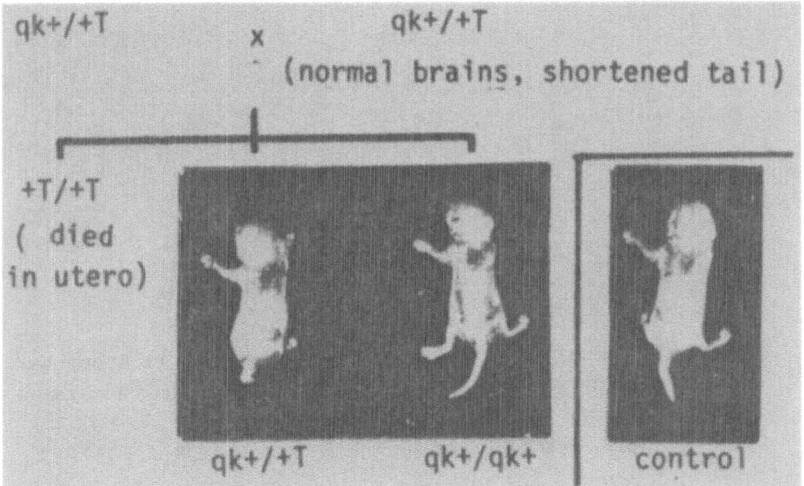

FIG. 8.   Newborn mice of homozygous ($qk+/qk+$) and heterozygous ($qk+/+T$) quaking were obtained by mating a pair of doubly heterozygous quaking and T ($qk+/+T$).

FIG. 9. Outgrowth from the explant of newborn homozygous quaking mutant mouse, 2 days in culture. No differences
were observed between the homozygous and heterozygous
cultures. A; Fibroblast-like cells. B; Immature cells, probably
astrocytes.

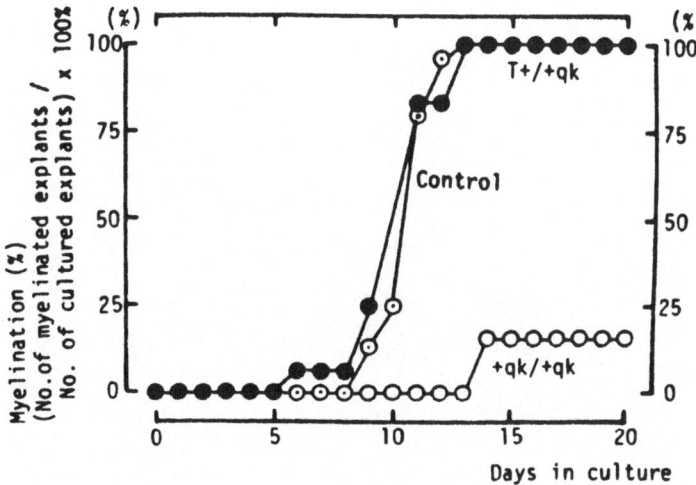

FIG. 10.   Rate of myelination in cultured cerebellum from
homozygous (○), heterozygous (●) and control (*ddY* mouse,
(◉). Myelination is expressed as a percentage of the number
of myelinated explants among the cultured explants.

vation, the explants from quaking mice on the 13th day of culture re-
presented only lipid-droplets and myelin-like fragments (Fig. 11). Quaking
explants, in all other respects examined by light microscopy, were as
healthy as the controls; flattening of the explants, emigration of macro-
phages, and organization of cavities lined with ependima displayed normal
development of neuronal cell bodies and axons, as judged in the living
state and after Bodian silver impregnation (Fig. 12). The oligodendrocyte
itself appears to be responsible for the dysmyelination.

When the conditioned medium from the quaking culture was added
to the control culture, it had no effect on myelination and the appearance
of astrocytes and other cells. Therefore, the possibility that some toxic
humoral factor caused the dysmyelination could be eliminated. But there
is still the possibility of a dilution effect or the short life of active factors
that cannot be eliminated.

Shiverer in vitro

Features of cultured cerebellum from shiverer mutant mice were similar
to that of quaking. Cultures from shiverer were observed to be either to-

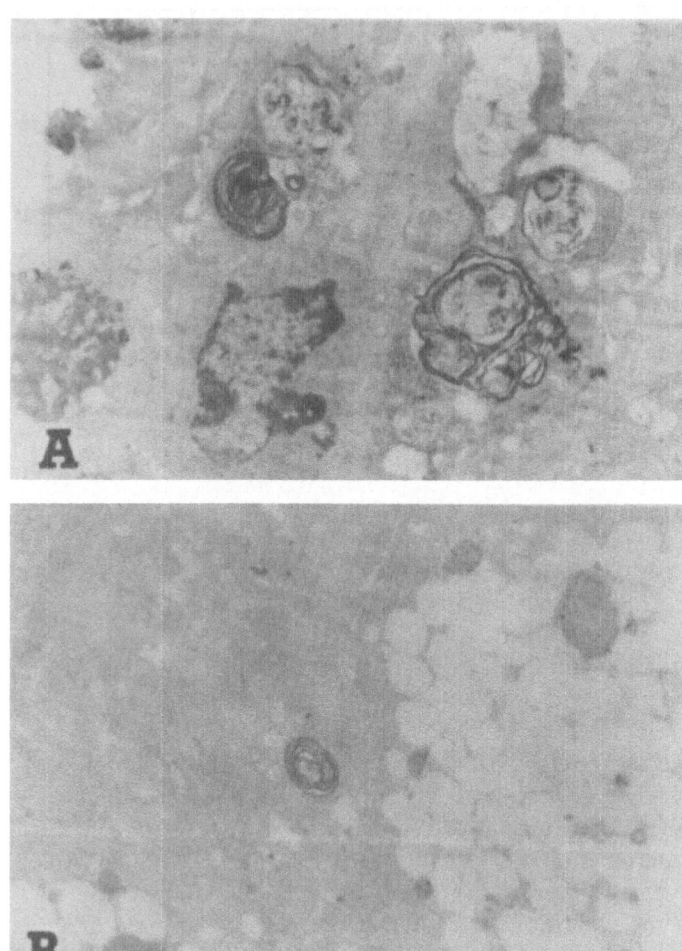

FIG. 11. Electronmicroscopic photograph of cultured quaking mutant mouse cerebellum. A; Myelin-like fragment. B; Lipid droplets.

tally devoid of myelin or slightly myelinated throughout the experiment (Fig. 13), but there was no remarkable difference in the abundance and shape of cells in outgrowth regions (Fig. 14) and neuronal cells (Fig. 15) with regard to Bodian silver impregnation and the living state.

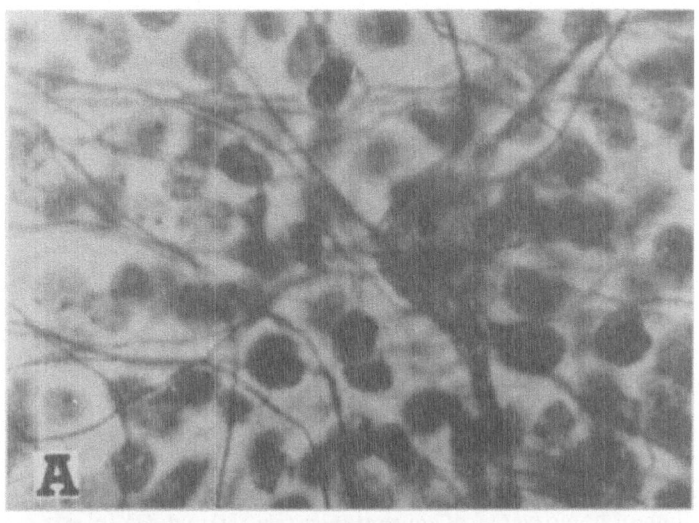

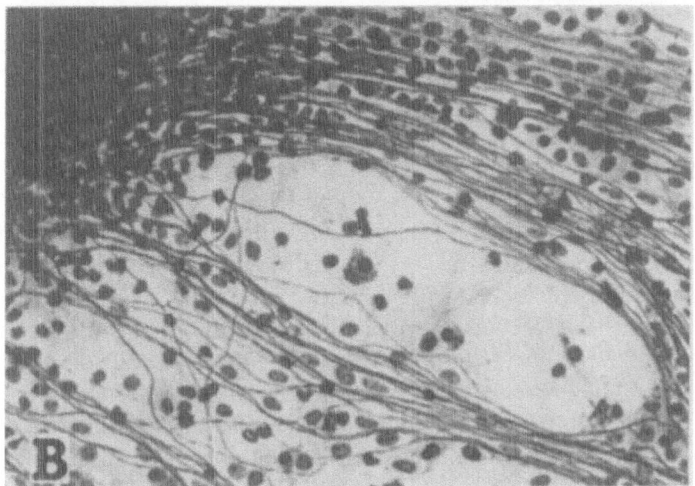

FIG. 12. Neuron (A) and axons (B) after Bodian silver impregnation ·in the culture of the quaking mouse. It was observed that abundant axons coursed throughout the cultured explants from cerebellum of homozygous quaking as well as heterozygous or *ddY* mice.

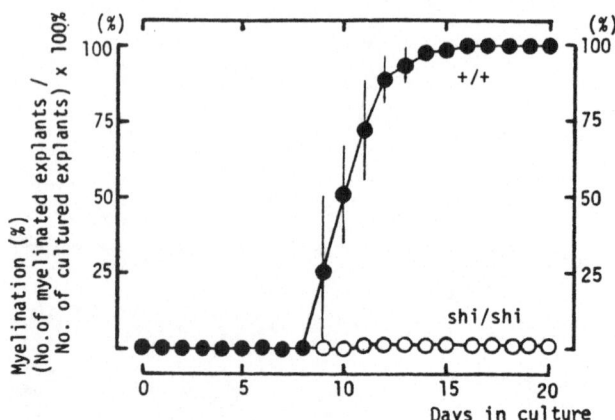

FIG. 13. Myelination in the organo-typic culture of the cerebellum of homozygous of shiverer mice (*shi/shi*) and *ddY* mice (+/+). No myelination was observed in the shiverer. Each point for myelination (%) represents the mean ± S.D. Homozygous males and females were used to produce the homozygote of shiverer.

Recently, it has been hypothesized that the processes of astrocytes interfere with contact between the axon and oligodendrocyte process resulting in dysmyelination in the jimpy mutant mouse.[37] By analyzing the shiverer mutant central nervous system, astrocyte hypertrophy was detected using immunohistochemical staining.

Since most of the astrocytes are divided after birth, an antimitotic drug was added to the medium to suppress the division of the cells soon after the incubation. They revealed that the addition of MAM did not have a stimulatory effect on myelination, although the cells that proliferated soon after birth, which were not connected with myelinogenesis, diminished in number. It was then assumed that dysmyelination in shiverers was due to the genetic defect of oligodendrocytes rather than to the hypertrophy of astrocytes.

Even though poor myelin formation was observed in the central nervous system of shiverer, the activity of CNPase, which is one of the myelin components,[22,38] was found to be as high as that of the control.[18] However, in jimpy and quaking the CNPase activities were found to be very low, reflecting poor myelin formation.[39,40] In the shiverer cerebellar

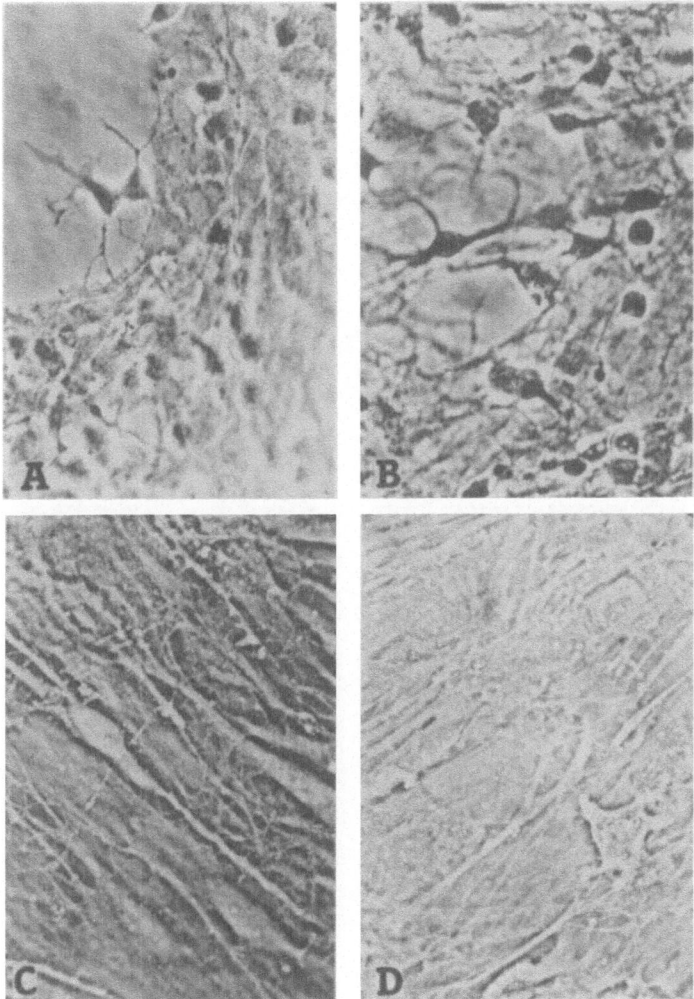

FIG. 14. Normal growth of glial cells in the cerebellar culture from shiverer mutant mice.

culture, the activity of CNPase was increased as in the control (Fig. 16).
Therefore, it seems that the CNPase molecules are formed normally in
shiverer both *in vivo* and *in vitro* in spite of the myelin deficiency. The
onset of the increase of CNPase activity in the shiverer almost coincided

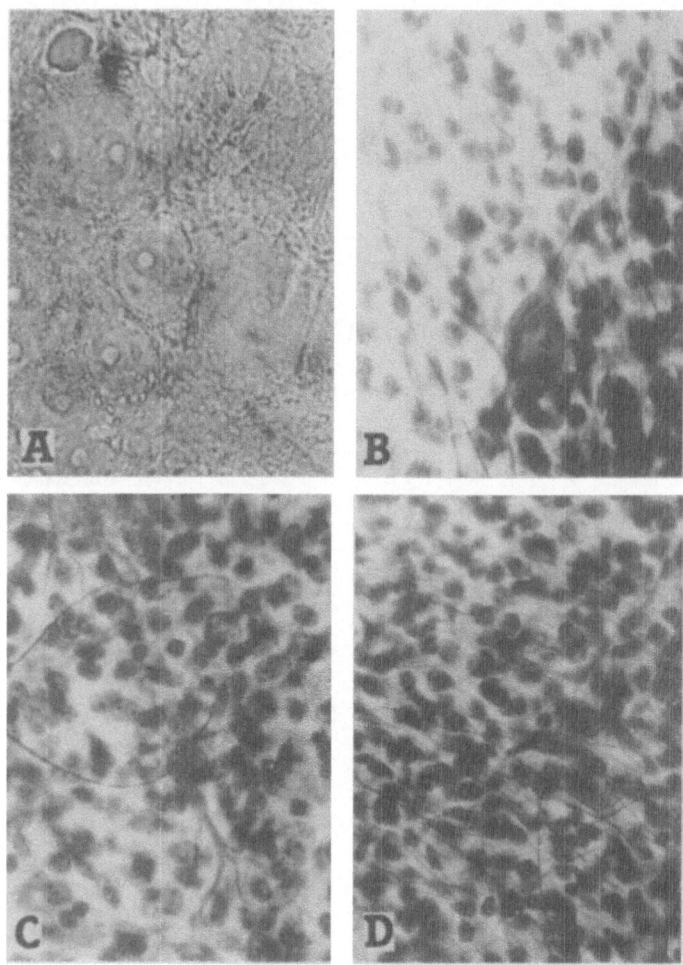

Fig. 15.   Neuron and axons in cultured shiverer. A: Living
state. B, C, D; Bodian silver impregnation.

with the morphological myelin formation in the control. The shiverer
mutant mouse appears to be a good model for investigating the factor
that controls the schedule of myelin assembly and the time schedule of
the synthesis of myelin components such as the CNPase molecule and
myelin basic protein.

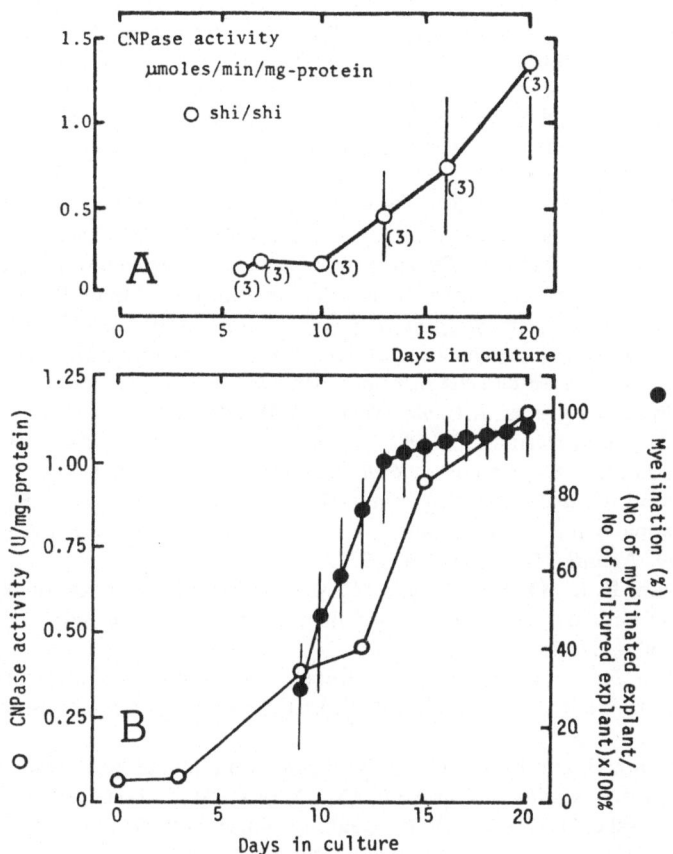

FIG. 16. Developmental change of CNPase activities in the cerebellar culture from shiverer (A) and control (B). Increase of CNPase activity was parallel to the appearance and accumulation of myelin in the control culture (●). Activity of CNPase in shiverer culture also increased with days of culture despite the lack of myelination.

REFERENCES

1. Peterson, E. R. and Murray, M. R.: Myelin sheath formation in cultures of avian spinal ganglia. *Am. J. Anat.*, **96**: 319–332, 1955.
2. Hild, W.: Myelogenesis in cultures of mammalian central nervous tissue. *Z. Zellforsch.*, **46**: 71–95, 1957.

3. Bornstein, M. B. and Murray, M. R.: Serial observations on patterns of growth, myelin formation, maintenance and degeneration in cultures of new-born rat and kitten cerebellum. *J. Biophys. Biochem. Cytol.*, **4**: 499–513, 1958.

4. Field, E. J., Hughes, D. and Raine, C. S.: Electron microscopic observations on the development of myelin cultures of neonatal rat cerebellum. *J. Neurol. Sci.*, **8**: 49–60, 1968.

5. Wolf, M. K.: Differentiation of neuronal types and synapses in myelinating cultures of mouse cerebellum. *J. Cell. Biol.*, **22**: 259–279, 1964.

6. Wolf, M. K. and Dubois-Dalcq, M.: Anatomy of cultured mouse cerebellum, I: Golgi and electron microscopic demonstrations of granulle cells, their afferent and efferent synapses. *J. Comp. Neurol.*, **140**: 261–280, 1970.

7. Wolf, M. K.: Anatomy of cultured mouse cerebellum, II: Organotypic migration of granule cells demonstrated by silver impregnation of normal and mutant cultures. *J. Comp. Neurol.*, **140**: 281–298, 1970.

8. Seil, F. J. and Herndon, R. M.: Cerebellar granule cells *in vitro*: A light and electron microscope study. *J. Cell. Biol.*, **45**: 212–220, 1970.

9. Allerand, C. D.: Pattern of neuronal differentiation in developing cultures of neonatal mouse cerebellum: A living and silver impregnation study. *J. Comp. Neurol.*, **142**: 167–204, 1971.

10. Sidman, R. L., Dickie, M. M. and Appel, S. H.: Mutant mice (quaking and jimpy) with deficient myelination in the central nervous system. *Science*, **144**: 309–311, 1964.

11. Sidman, R. L., Green, M. C. and Appel, S. H.: Catalog of the Neurological Mutants of the Mouse. Harvard University Press, Cambridge, Massachusetts, 1965.

12. Watanabe, I. and Bingle, G. J.: Dysmyelination in quaking mouse: Electron microscopic study. *J. Neuropath. Exp. Neurol.*, **31**: 352–369, 1972.

13. Wisniewske, H. and Morell, P.: Quaking mouse; Ultrastructural evidence for arrest of myelinogenesis. *Brain Res.*, **29**: 63–73, 1971.

14. Mikoshiba, K., Nagaike, K., Aoki, E. and Tsukada, Y.: Biochemical and immunohistochemical studies on dysmyelination of quaking mutant mice *in vivo* and *in vitro*. *Brain Res.*, **177**: 287–299, 1979.

15. Bidde, R., March, E. and Miller, J. R.: *Mouse News letter*, **48**: 24, 1973.

16. Chernoff, G., March, E. and Miller, J. R.: *Mouse News Letter*, **5**: 12, 1974.

17. Duponey, P., Jacque, C., Bourre, J. M., Cesselin, J., Privat, A. and Baumann, N.: Immunochemical studies of myelin basic protein in shiverer mouse devoid of major dense line of myelin. *Neuroscience Letter*, **12**: 113–118, 1979.

18. Mikoshiba, K., Aoki, E. and Tsukada, Y.: 2′,3′-cyclic nucleotide 3′-phosphohydrolase activity in the central nervous system of a myelin deficient mutant (shiverer). *Brain Res.*, **192**: 195–201, 1980.

19. Mikoshiba, K., Nagaike, K. and Tsukada, Y.: Subcellular distribution and developmental change of 2′,3′-cyclic nucleotide 3′-phosphohydrolase in

the central nervous system of the myelin-deficient shiverer mutant mice. *J. Neurochem.*, **35**: 465–470, 1980.

20. Mikoshiba, K., Nagaike, K., Takamatsu, K. and Tsukada, Y.: Developmental change of 2′,3′-cyclic nucleotide 3′-phosphohydrolase activity in the nervous system of the shiverer mutant mice *in vivo* and *in vitro*. In: Neurological Mutants Affecting Myelination. INSERM Symposium No. 14 (ed. N. Baumann), pp. 349–354. Elsevier/North-Holland, Amsterdam, 1980.

21. Bornstein, M. B.: Reconstituted rat-tail collagen used as substrate for tissue culture on coverslips in Maximow slides and roller tubes. *Lab. Invest.*, **7**: 134–137, 1958.

22. Kurihara, T. and Tsukada, Y.: The regional and subcellular distribution of 2′,3′-cyclic nucleotide 3′-phosphohydrolase in the central nervous system. *J. Neurochem.*, **14**: 1167–1174, 1967.

23. Kissan, J. M. and Robins, E.: The fluorometric measurement of deoxyribonucleic acid in animal tissues with special reference to the central nervous system. *J. Biol. Chem.*, **233**: 184–188, 1958.

24. Lowry, O. H., Rosebrough, N. J., Farr, A. L. and Randall, R. J.: Protein measurement with the Folin phenol reagent. *J. Biol. Chem.*, **193**: 265–275, 1951.

25. Allerand, C. D. and Murray, M. R.: Myelin formation *in vitro*: Endogeneous influences on cultures of newborn mouse cerebellum. *Arch. Neurol.*, **19**: 292–301, 1968.

26. Kies, M. W., Driscoll, B. F., Seil, F. J. and Alrord, E. C.: Myelination inhibition factor; Dissociation from induction of experimental allergic encephalomyelitis. *Science.*, **179**: 689–690, 1973.

27. Altman, J.: Autoradiographic and histological studies of postnatal neurogenesis, II: A longitudinal investigation of the kinetics, migration and transformation of cells incorporating tritiated thymidine in infant rat, with special reference to postnatal neurogenesis in some brain regions. *J. Comp. Neurol.*, **128**: 431–474, 1976.

28. Altman, J.: Autoradiographic and histological studies of postnatal neurogenesis, III: Dating the time of production and onset of differentiation of cerebellar microneurons in rats. *J. Comp. Neurol.*, **136**: 269–294, 1969.

29. Fujita, S.: Analysis of neuron differentiation in the central nervous system by tritiated thymidine autoradiography. *J. Comp. Neurol.*, **122**: 311–328, 1964.

30. Fujita, S., Simada, M. and Nakamura, T.: ³H-thymidine autoradiographic studies on the cell proliferation and differentiation in the external and the internal granular layers of the mouse cerebellum. *J. Comp. Neurol.*, **128**: 191–208, 1966.

31. Privat, A. and Drian, M. J.: Postnatal maturation of rat Purkinje cells cultivated in the absence of two afferent systems: An ultrastructural study. *J. Comp. Neurol.*, **166**: 201–244, 1976.

32. Silberberg, D. H., Dorfman, S. H., Latoritzki, N. and Younkin, L. H.: Oligodendrocyte differentiation in myelinating cultures. In: Tissue Culture in Neurobiology (ed. E. Giacobin), pp. 409–500. Raven Press, New York, 1980.

33. Dorfman, S. H., Holtzer, H. and Silberberg, D. H.: Effect of 5-bromo-2'-deoxyuridine or cytosine-B-D-arabino furanoside hydrochloride on myelination in newborn rat cerebellum cultures following removal of myelination inhibiting antiserum to whole cord or cerebroside. *Brain Res.*, **104**: 283–294, 1976.

34. Skoff, R. P., Price, D. L. and Stocks, A.: Electron microscopic autoradiographic studies of gliogenesis in rat optic nerve. I. Cell proliferation. *J. Comp. Neurol.*, **169**: 291–312, 1976.

35. Skoff, R. P., Price, D. L. and Stocks, A.: Electron microscopic autoradiographic studies of gliogenesis in rat optic nerve. II. Time of origin. *J. Comp. Neurol.*, **169**: 313–334, 1976.

36. Mares, V. and Bruckner, G.: Postnatal formation of non-neuronal cells in the rat occipital cerebrum; An autoradiographic study of the time and space pattern of cell division. *J. Comp. Neurol.*, **177**, 519–528, 1978.

37. Skoff, R. P.: Myelinodeficient in the jimpy mouse may be due to cellular abnormalities in astroglia. *Nature* (London), **264**: 560–562, 1976.

38. Tsukada, Y. and Suda, H.: Solublization and purification of 2',3'-cyclic nucleotide 3'-phosphohydrolase (CNP-A) from bovine cerebral white matter. *Cell & Moll. Biol.*, **26**: 493–504, 1980.

39. Kurihara, T., Nussbaum, J. L. and Mandel, P.: 2',3'-cyclic nucleotide 3'-phosphohydrolase in the brain of the "Jimpy" mouse, a mutant with deficient myelination. *Brain Res.*, **13**: 401–403, 1969.

40. Kurihara, T., Nussbaum, J. L. and Mandel, P.: 2',3'-cyclic nucleotide 3'-phosphohydrolase in the brains of mutant mice with deficient myelination. *J. Neurochem.*, **17**: 993–997, 1970.

General Comments

I wish to express my views from the standpoint of a neurobiologist. It is well known that the development of the nervous system is governed by genetic and environmental factors, but it is important to confirm their biological limits. More research is necessary in the field of molecular biology concerning the genetic control of the process of organogenesis and differentiation of the nervous system. I think that research in this direction will provide new breakthroughs in neurobiology. This is one of the reasons why early development is one of the subjects of this symposium, although this point was not very well clarified. I believe it is necessary to fill the gap between early development and the appearance of mutant mice.

On more aspect of the genetic approach is the investigation of the relation between genetic defects and abnormalities in the development of the nervous system, using neurological mutants. Such studies have made major progress by producing and analyzing chimeras. I hope that this kind of research will proceed even further in the future. There are also many studies that attempt to examine the physiological and pathological background of the nervous system using mutants as experimental models. In this case the mutant is only a tool, and this cannot be called a genetic approach. Such studies nevertheless constitute valuable medical research in that they are useful as disease models for the development of therapeutic methods. I believe the mainstream of the genetic approach should be composed of studies that help to clarify the role of the gene in the development and differentiation of the nervous system. We must continue our efforts to find an experimental system that will meet such an objective.

Yasuzo Tsukada, M.D.

Executive Members of Japan Medical Research Foundation